PEST CONTROL: An Assessment of Present and Alternative Technologies

VOLUME IV

Forest Pest Control

The Report of the Forest Study Team
Study on Problems of Pest Control
Environmental Studies Board
National Research Council

NATIONAL ACADEMY OF SCIENCES
Washington, D.C. 1975

This is Volume IV of *Pest Control: An Assessment of Present and Alternative Technologies*. The other volumes are:

Volume I	*Contemporary Pest Control Practices and Prospects: The Report of the Executive Committee*
Volume II	*Corn/Soybeans Pest Control*
Volume III	*Cotton Pest Control*
Volume V	*Pest Control and Public Health*

NOTICE: The project that is the subject of this report was approved by the Governing Board of the National Research Council, whose members are drawn from the Councils of the National Academy of Sciences, the National Academy of Engineering, and the Institute of Medicine. The members of the Committee responsible for the report were chosen for their special competences and with regard for appropriate balance.

This report has been reviewed by a group other than the authors according to procedures approved by a Report Review Committee consisting of members of the National Academy of Sciences, the National Academy of Engineering, and the Institute of Medicine.

Library of Congress Catalog Card No. 75-37297

International Standard Book No. 0-309-02413-7 (this volume)
0-309-02409-9 (entire set)

Available from

Printing and Publishing Office
National Academy of Sciences
2101 Constitution Avenue, N.W.
Washington, D.C. 20418

Printed in the United States of America

PREFACE

This study originated in discussions in the Executive Offices of the National Academy of Sciences in March 1971. The Environmental Studies Board (ESB)* subsequently continued the discussions and appointed a Planning Committee to develop a detailed proposal. The Planning Committee, sponsored by funds from the Scaife Family Charitable Trusts, submitted its report to the ESB in April 1972. The ESB approved the report and appointed Dr. Donald Kennedy of Stanford University as Chairman of the Executive Committee of the Study on Problems of Pest Control: A Technology Assessment. Financial support for the study was secured from the U.S. Department of Agriculture, the U.S. Environmental Protection Agency, and the Ford Foundation. We thank these public and private agencies for their assistance.

The Executive Committee was charged to assess the diverse effects on society that arise from efforts to control pests, and to evaluate long-range alternatives

---

*The National Academy of Sciences complex (National Academy of Sciences, National Academy of Engineering, Institute of Medicine) functions as official but independent advisor to the federal government in matters relating to science and technology through its operating agency, the National Research Council. The National Research Council comprises four Assemblies and four Commissions each of which deals with specific disciplines or societal problems. The Commission on Natural Resources, formed in 1973, oversees the work of five boards, among them the Environmental Studies Board. Thus the pest control study, having originated as described above, completed its activities under the aegis of the Commission on Natural Resources.

available for accomplishing pest control. The charge also called explicitly for an examination of social and institutional factors important to the conduct of pest control practices, factors that must be considered in any recommendations for changes in pest control policy. The Planning Committee and the Executive Committee concluded that the most effective way to examine pest control problems would be through a series of case studies. Four such studies were selected. Three of them centered on specific agroecosystems: cotton, corn and soybeans, and timber. The fourth case study dealt with issues related to the control of pests of public health significance in the United States and abroad. Each case study examined the pest species involved, current control measures, and the benefits and costs of controlling the pests by various techniques.

This volume comprises the report of the members of the Forest Study Team. Although the team's task concerns just one aspect of the general problem of controlling pest organisms, it deals with a highly diverse aggregate of environments and organisms. The broadleaved forests of wet bottomlands in the Mississippi Delta, the old-field white pine forests of New England, and the dense coniferous forests of coastal Oregon, for example, have certain structural features in common but differ entirely in the plant and animal species that compose them and in the ways that these species react to man's treatments, past and future.

A discussion of pest control practices in forests must take into account not only the geographical and biological differences in forest types and pest organisms but also the different management objectives of various owners and beneficiaries of the nation's forests. Modern society seeks a variety of sometimes conflicting "goods and services" from forested lands--wood, water, stability of steep and fragile soils, grazing for domestic animals, habitat for wildlife, and a great range of recreational, scenic, and aesthetic experiences. Every forest provides several of these, and some provide all of them. Part of the forest setting is maintained or recreated in much modified form in the suburban and urban landscape. The priorities assigned to these many values by landowners or the concerned public influence the way that forests are treated.

The task of the Forest Study Team, then, has been quite different from that of the commodity teams. This difference is reflected in the report that follows: the spectrum of management interests is wide; there are many kinds of trees, and a great many organisms that are pests in some circumstances and in some forests. We have restricted our attention in this report to major pests of live trees, omitting sources of damage to roundwood and wood products. Similarly, we have not considered the damage that man himself as a forest user may do to the land and trees, except where his actions have created or exacerbated other pest problems. The pests discussed fall into four major classes: microorganisms, insects, vertebrates, and weeds. Omission or only brief mention of many of the minor invertebrate pests, birds, and other pests does not mean that their importance to some trees or at some time is not recognized; rather, it reflects the limits of space and of our knowledge of examples.

The team met five times between June 1973 and April 1974 and also went into the field on two occasions. The first was in Seaside, Oregon, where we toured a portion of Crown-Zellerbach's land to view firsthand some of the industrial forestry practices of the Pacific Northwest. The second field trip included an aerial view of the southern pine beetle outbreak in the Athens, Georgia, area, as well as a tour of a Georgia-Pacific plywood mill and a look at southern industrial management practices on Georgia Kraft Company land. Committee members prepared working papers on selected topics between the meetings; this volume is based primarily on these papers and the team's discussions.

The reports of the Executive Committee and the study teams are all available from the National Academy of Sciences in five volumes under the general title *Pest Control: An Assessment of Present and Alternative Technologies*. Volume I comprises the report of the Executive Committee and all the summaries and recommendations of the four study team reports. The report of the Executive Committee, in addition to drawing on the findings of the individual study teams, covers other areas not addressed by those teams. The remaining four volumes, bound separately, are as follows: Volume II: *Corn/Soybean Pest Control*; Volume III: *Cotton Pest Control*; Volume IV: *Forest Pest Control*; and Volume V: *Pest Control and Public Health*.

ACKNOWLEDGMENTS

The members of the study team would like to express our sincere appreciation to those who contributed to our deliberations, by attending our meetings at our invitation and presenting their thoughts on selected topics, by arranging portions of the team's two field trips and serving as our guides, or by providing special material as background information. The following deserve special thanks for their assistance:

Paul Lauterbach, Weyerhaeuser Corporation; Donald Malmberg, Crown-Zellerbach Corporation; Charles Sartwell, U.S. Forest Service, Corvallis, Oregon; Carl Smith, administrator of the Oregon Forest Practices Act; Robert Gustafson, John Huber, and Richard Zoelch, administrators of pest control operations on urban trees in and around Portland, Oregon; Amel Landgraf, U.S. Forest Service, Atlanta, Georgia; Warren Carson, Georgia Kraft Company; Paul Butts, Frank Craven, Theron Devereaux, Rip Fontane, John Hammond, and W. D. Williams, of the Georgia Forestry Commission; John Chansler, U.S. Forest Service, Washington, D.C.; Neil MacGregor, U.S. Forest Service, San Francisco, California; and William Waters, U.S. Forest Service, Berkeley, California.

Judith Hough, of the National Research Council staff, attended all meetings of the study team, assisted in arrangements, and participated in the preparation of the final report. Dr. John Perkins served as Principal Staff Officer for the study as a whole, and Bette Fields took over his responsibilities in the final phases of the study. Linda Jones, Lucette Comer, Mary Dong, Susan Lesser, Christina Neumann, and Estelle Miller provided able and tireless secretarial support. We are grateful for their generous and effective contributions.

FOREST STUDY TEAM

Richard S. Miller, *Chairman*
Professor
School of Forestry and Environmental Studies
Yale University
New Haven, Connecticut

Donald L. Dahlsten
Professor of Entomology
University of California
Berkeley, California

Fred B. Knight
Director
School of Forest Resources
Nutting Hall, U.M.O.
University of Maine
Orono, Maine

Michael Newton
Associate Professor
School of Forestry
Oregon State University
Corvallis, Oregon

William A. Niering
Professor of Botany
Department of Botany
Connecticut College
New London, Connecticut

David E. Pesonen
Garry, Dreyfus, McTernan, Botsky, Herndon & Pesonen, Incorporated
Attorneys at Law
San Francisco, California

William H. Smith
Assistant Dean and Associate Professor of Forest Pathology
School of Forestry and Environmental Studies
New Haven, Connecticut

Earl L. Stone
Charles Lothrop Pack Professor of Forest Soils
Cornell University
Ithaca, New York

Henry J. Vaux
Department of Forestry and Conservation
University of California
Berkeley, California

CONTENTS

SUMMARY AND RECOMMENDATIONS

The "products" provided by a forest depend upon the objectives of the managers and users of the land. In this sense, forests do not provide us with only one well-defined product, nor is there any single set of pests or any one organism that is a pest in all forest situations. Also, the future of a forest with even a brief rotation from seedlings to mature or harvestable trees cannot be decided the year it is planted, if only because of shifting and unpredictable future values of its potential products.

Thus, the existence of many different objectives for different forest lands, or for a single forest over time, creates a situation in which it is important that managers avoid irreversible pest control decisions that might have unwanted effects in the future. This philosophy also suggests that pest control should not be aimed solely at killing pest organisms and should be undertaken only when the activity of a pest can be clearly shown to interfere significantly with management objectives. This also suggests the need for careful cost/benefit analyses that ensure that those objectives will be served without undue cost or loss of other important benefits. With these general statements as a philosophical background, the Forest Study Team offers the following recommendations.

In the area of pest control research, priorities have often been set under political pressure for immediate answers, with too much regard for specific short-term problems and too little consideration for broader management objectives. Part of this problem arises from portions of the Forest Pest Control Act of 1947 that define policy, and create a division between the administrative and research arms of the U.S. Forest Service (USFS). The Act also tends to promote an "action" attitude that may inhibit solutions other than short-term, direct chemical control.

We, therefore, recommend that the Forest Pest Control Act be reexamined with a view toward repeal of its policy sections or replacement with a new act that is consistent with the management philosophy presented in this report.

Destructive harvesting techniques and harvesting without reforestation, where forest lands have been exploited for immediate profit, create "weed" problems and loss of future productivity. Reforestation laws that require that specified stocking standards be met within a certain number of years after harvest appear to be functioning effectively in Oregon and Washington and could be effective in other regions where even-age management is practiced.

We recommend adoption of similar laws by other states where such a law would be appropriate.

The impact of pests is part of the USFS inventory, but present methods are reasonably good for insects, poor for diseases, and almost nonexistent for weeds. There is a general need for better information and better inventory data on a variety of subjects related to forest management, and it is particularly important that loss estimates be made in terms of the whole spectrum of social values involved and that they be consistent with all of the different management objectives involved.

We recommend that the USFS be encouraged, and provided with the necessary funds, to increase its efforts in the development of accurate methods of estimation and fuller appraisals of losses due to pests.

Detection of insect (and other pest) damage is still a major problem in forest pest management. At present, detection of insect damage is almost always based on visible damage, often by air survey. The results are often obtained so late that treatment is applied to already declining insect populations, and it is impossible in such cases to evaluate the effectiveness of control measures. With more adequate detection and more carefully controlled evaluation of control efforts, it would be possible to reduce or eliminate prophylactic treatments for insect control, which are used in many urban and suburban forests, when there is no genuine evidence of significant hazard from insect pests.

We recommend that sampling procedures be improved and better methods developed; that control projects include the cost of accurate evaluation procedures, including the establishment of adequate comparisons between treated and untreated areas; and that evaluation of control methods include effects on nontarget organisms and natural enemies of the target insect.

Studies are needed on all aspects of silvicultural control, including site preparation and species selection, cultural practices after the establishment of a stand (such as thinning, including the use of herbicides), and postharvest sanitation and salvage techniques. Where ecological, silvicultural, and economic considerations permit, pest management should utilize cultural procedures. Research on silvicultural methods is especially needed in the protection of forests from bark beetle infestations, and in forest manipulation to avoid spruce budworm damage. Certain cultural techniques, such as sanitation/salvage, require political and economic coordination as well as technical skill. The case of the Southern pine beetle, for example, demonstrates the need for careful organization of procedures and development of an adequate labor force, transportation facilities, and industrial capacity for rapid salvage in the event of an insect epidemic.

We recommend expanded research on the development and use of silvicultural methods of forest pest control.

The use of more than one method to prevent damage by pests is sometimes referred to as "integrated control" or, more recently, "integrated pest management." The term also implies maximum use of natural controls of pest populations wherever feasible. The concept of integrated pest management is also relevant to vegetation management, as in the case of selective removal of unwanted trees on rights-of-way and establishment of low-growing shrub cover.

We recommend that research on forest pest control be closely coordinated with overall research on silviculture and forest management, to better achieve the objectives of integrated pest management.

The use of biotic agents to control forest pests ("biological control") is still in an early stage of research and development. With respect to insects, the greatest success has been against introduced pests, and this is also the area that shows the greatest promise. However, biological control of tree disease also shows promise. At present, the only practical use of a biotic agent to reduce the activities of pathological organisms is the inoculation of pine stumps with the fungus *Peniophora gigantea* against infection by *Fomes annosus*, but potential exists to develop biological systems antagonistic to wood decay fungi, damping-off fungi, canker and vascular disease fungi, and dwarf mistletoe.

The use of insect pheromones to attract pests to traps, or to disrupt their normal activities and thereby reduce their impact as pests, is being studied, and some pheromones have been used in preliminary control studies, but the only proven use for insect pheromones in forest pest control, to date, is as a survey tool.

Another potential form of biological control is species mixtures, in which pest problems are reduced by maintaining diversity of species. For example, there is a significantly higher incidence of *Poria weirii* root infection in pure stands of Douglas fir than in mixed stands that include red alder.

We recommend increased support of research on these and other promising methods of biological control.

Forestry has depended almost entirely upon pesticides developed for use in agriculture, because it is in the "minor use" category with respect to the pesticide market. There is little incentive for manufacturers to develop and register formulations for minor uses, and useful, relatively safe pesticides have been taken off the market when their use in agriculture declined. The result is that many pesticides that are potentially useful in forestry are unavailable, and in some cases, there have been no alternative solutions to pest problems, while in others the alternative has been more costly with more adverse environmental effects.

We recommend an examination of mechanisms that would provide a greater range of chemicals for the specific needs of minor users.

Incentive programs to improve forests on small private lands and to support sound pest management programs have had limited success, partly because the recipients of incentives have largely been landowners who would practice good forest management in any event, rather than marginal operators whose stand management is contingent upon receiving education as well as benefits.

We endorse recommendation Number 10 of the President's Advisory Panel on Timber and the Environment (April 1973), which calls for the federal government to maintain and improve incentive programs through development of income tax incentives, advice and service to forest owners and associations, and cost-sharing for intensive forest management practices.

The flow of information and recommendations about pest problems in urban, suburban, and other intensively managed areas needs to be carefully examined. Pest control decisions are frequently made by nonprofessionals, such as homeowners, on the advice of other nonprofessional or semiprofessional "experts," or pesticides are purchased and applied with no advice whatever. Cities and towns need more trained practitioners of plant welfare with practical expertise in ecology, physiology, entomology, pathology, and soils.

We recommend the training and appointment of such professionals in more urban and suburban areas and a vigorous program of public education on the use and nonuse of pesticides.

1

INTRODUCTION TO AN ANALYSIS OF PEST CONTROL PRACTICES IN U.S. FORESTS

OVERVIEW

One out of every three acres in the United States is classified as forest land. The 750 million acres in forests would cover the United States east of the Mississippi River with enough left over to carpet Texas and part of California. In addition, forest trees are important features of many urban and suburban areas that are not classified as forest land. There are about 585 separate species of trees native to the United States, and, in addition, more than 90 foreign species have become naturalized here (Little 1949). American forests are thus both of vast extent and great biological variety.

In every forest there are a large number of other organisms--animals (including insects), microorganisms, shrubs, and annual and perennial herbs--living in intimate ecological relationship with the trees. These relationships may be favorable, inimical, or essentially neutral to the survival and growth of the trees, depending on the specific forest situations. From the biological point of view, all of these organisms--trees, mammals, birds, insects, microbes, and secondary vegetation--fulfill characteristic ecological roles. Each is an integral part of the natural forests, and any substantial natural or man-induced change in the population of one organism is likely to have ecologically significant effects on one or more of the others.

About 165 of the native and introduced tree species are recognized as having major actual or potential importance to man, as sources of wood products, food, or medicine; as aesthetic features of the landscape; or as essential protective cover. As for the associated organisms, some may contribute directly to human welfare; game animals, flowering shrubs, birds, and bees are obvious examples.

These same organisms may, however, have adverse effects in some situations. Bears may severely damage pole-sized Douglas firs and redwoods by stripping off bark; flowering currant is an alternative host for blister rust, which is lethal to young white pine trees; and a bees' nest may be a significant human hazard if located in a heavily used campground. The trees themselves may be "pests" if they obscure a cherished view, contribute unduly to fire or windstorm hazards to human habitation, or draw excessive quantities of moisture from a critical watershed.

Pests are organisms that diminish the value of resources in which man is interested. An organism can be classed as a forest pest only on the basis of a recognized set of forest management objectives and a clear understanding of the organism's functions as an element in the particular ecosystem of which it happens to be a part.

This view of the forest pest problem differs in major degree, if not in kind, from the view of pests that is characteristic in crop agriculture and public health. In both those areas of concern, man's objectives are usually more single-minded than is the case in forestry. On public forest land, for example, the simultaneous existence of more than one valid objective for a single area is specifically recognized by the Multiple-Use Sustained Yield Act of 1960. On any forest site, the question of what management objectives are appropriate to the particular area becomes a central feature of any discussion of the forest pest situation.

Table 1-1 suggests this diversity in management objectives by indicating the current ownership status of commercial forest land in the United States. Forest land owned by private industry, which accounts for only 13.5 percent of the commercial forest land, is managed primarily for the profitable production of timber as a commodity for conversion into wood products. In national forests and other forests owned by the public, however, such commodity production is usually only one of a number of explicit forest management objectives, embraced within the broad concept of "multiple use." Here the relative priority of production of timber commodities and other forest uses such as providing sites for outdoor recreation or protecting watersheds is not clearly established on large portions of the area involved. On the "other private" category of ownership (almost 60 percent of the forest land) we know that management objectives vary greatly among the 3 or 4 million individual owners, but there is little information that would permit us to identify ownership objectives on any one particular property.

TABLE 1-1 Commercial Forest Land in the United States by Ownership and by Region, January 1, 1970 (million acres)[a]

Region	Total	National Forest	Other Public	Private Industry	Other Private
North	177.9	10.5	21.4	17.6	128.4
South	192.5	10.8	6.5	35.3	140.0
Rocky Mountains	61.6	38.8	7.2	2.2	12.5
Pacific Coast	67.6	30.9	9.1	12.2	15.5
TOTAL	499.7	92.0	44.2	67.3	296.4

[a] Totals may not add because of rounding.

SOURCE: USFS 1972, Table 2.

This diversity of objectives makes it virtually impossible to characterize any organism categorically as a forest pest. Instead it forces us to consider forest pest problems within some explicit forest management framework, where the objectives of management are known, the significant ecological variables can be quantified, and management capabilities can be evaluated in relation to possible alternative pest control strategies.

The biological variability inherent in forest pest problems has already been noted. Most forest pests either favor certain species of trees over others or are completely specific with respect to the kinds of trees they will attack. The conditions that favor epidemic levels of damage vary greatly with different forest sites, stages of timber stand development, topographic situations, and seasonal weather patterns, to name only a few important sources of variation. As an example, Table 1-2 suggests something of this variation of the Pacific Coast forest in species composition, site quality, and age of stand. Similar variability in factors influencing susceptibility to pest damage is characteristic of each of the other regions.

Keeping in mind both the managerial and ecological dimensions of the complexity of the forest pest problem, we now consider a number of factors that are currently shaping forest resource use in the United States and that seem likely to continue exerting strong influence on forest management--and therefore, on the nature of forest pest problems--over the next several decades.

FOREST AREA

Following the first significant white settlement in what is now the United States, the total area of forest land dwindled fairly steadily as much of the existing forest was converted into crop land and related nonforest uses. This process of diminution of the forest base continued until the end of World War I. By 1900, however, significant amounts of previously cultivated land were being abandoned in the Eastern United States, both in the North and in the South. Largely through natural processes, most of this abandoned farm land reverted to forest during the ensuing decades. From early in the century to early in the 1950s, these accretions more than offset continued conversions of forests to other uses, so that for almost 50 years the area of land in forests appeared to increase modestly. That trend has now been reversed; total forest

TABLE 1-2 Commercial Forest Land in the Pacific Coast Region by Vegetation Type, Site Quality, and Stand Age, January 1, 1970 (million acres)[a]

Stand-size[b] Site Quality[c] Classes	Non-stocked	Douglas Fir	Ponderosa Pine	Hemlock Spruce	Fir Spruce	Other Softwood	Hardwood	All Types
Sawtimber								
120+	--	5.6	1.5	3.7	1.9	0.7	2.4	15.8
85-120	--	2.7	2.8	2.4	1.7	0.4	1.2	11.2
50-85	--	4.7	5.0	1.4	2.4	1.0	1.1	15.7
20-50	--	0.7	1.7	0.2	0.6	0.2	0.2	3.6
Poletimber								
120+	--	0.8	0.1	0.4	0.2	0.1	0.7	2.3
85-120	--	0.3	0.3	0.1	0.2	0.6	0.3	1.8
50-85	--	1.0	0.7	0.1	0.3	1.0	0.3	3.4
20-50	--	0.1	0.4	negl.	0.1	0.1	0.1	0.8
Saplings & Seedlings								
120+	1.5	1.4	0.1	1.0	0.2	0.1	1.1	5.4
85-120	0.7	0.6	0.3	0.3	0.2	0.2	0.4	2.7
50-85	1.1	0.8	0.4	0.3	0.2	0.4	0.6	3.7
20-50	0.4	0.2	0.2	negl.	negl.	0.1	0.2	1.1

TABLE 1-2 (Continued)

Stand-Size[b] Site Quality[c] Classes	Non-stocked	Douglas Fir	Ponderosa Pine	Hemlock Spruce	Fir Spruce	Other Softwood	Hardwood	All Types
All Size Classes								
120+	1.5	7.8	1.7	5.1	2.3	0.9	4.2	23.6
85-120	0.7	3.5	3.4	2.8	2.1	1.2	1.9	15.7
50-85	1.1	6.5	6.1	1.8	2.9	2.4	2.0	22.8
TOTAL	3.7	18.9	13.5	9.9	8.0	5.0	8.5	67.6

[a]Column and row totals may not add because of rounding.
[b]Stand-size classes are based on the predominant size of timber present, i.e., whether most of the trees are sawtimber ($\geq$11-in. diameter for hardwoods and for softwoods in the Pacific Coast region), poletimber ($\geq$5-in. diameter), saplings (1-5 in.) or seedlings (less than 1 in.).
[c]The site quality classification refers to potential growth in cubic feet of a fully stocked natural stand.

SOURCE: USFS 1972, Table 44.

area in 1970 was about 1.7 percent less than it had been in 1962.

This relatively fixed total forest base, however, is under steadily increasing human pressure. Burgeoning public interest in recreation is resulting in steady enlargement of the forest area reserved for recreation and park purposes. Each year, more forest is devoted to second homes, highways and other rights-of-way, or suburban development. The demand for forest products and thus for land on which to grow future timber supplies continues to increase. The open forest frontier that existed as recently as the 1940s has disappeared from the contiguous 48 states.

This is clearly shown in the trend of the USFS estimates of commercial forest land area, which was only 499.7 million acres in 1970, down 1.7 percent from the area in 1962. ("Commercial forest land" is land capable of growing continuous crops of industrial timber and not withdrawn for other forest uses. The capability referred to is essentially biological. Significant areas of commercial forest land appear to be of such low productivity that they are unlikely to support profitable timber growing under either existing or anticipated economic conditions [President's Advisory Panel on Timber and the Environment (PAPTE) 1973, Vaux 1973].) In the West, the decline in commercial forest area has resulted mainly from reservation of additional forest to meet increased demands for wilderness and recreation areas. In the South, considerable forest has been cleared for soybeans, other agricultural crops, or pasture. The land use pressures that have produced these changes will almost certainly continue as the nation tries to meet the diversified demands of an expanding population with an essentially fixed land-resource base (USFS 1973).

MAJOR FOREST LAND USES

The major uses of forest land include provision of habitat for wildlife, provision of an environment for diverse kinds of outdoor recreation, production of range forage for domestic livestock, protection of soil, protection and improvement of watersheds, growing and harvesting of timber, and preservation of rare or unique natural ecological or scenic features. Preservation of unique features and provision of environment for outdoor recreation were the uses first recognized by federal policy, with passage of the Yellowstone Park Act (1872). Subsequently, recognition was given to timber production and watershed protection as the basis for reservation of national forests (1897). Since 1905, the importance of all the forest uses enumerated above has been recognized in a wide variety of federal and state legislation.

Forests have an unusual capability to accommodate use for several of the above purposes at the same time (e.g., soil and watershed protection, preservation of scenic features, and provision of wildlife habitat on a single forest area; or timber growing and certain types of recreation on the same area, at least during most of the timber growth cycle). This "multiple use" capability is recognized as the appropriate means for achieving management goals on much publicly owned land, and to some degree it is a feature of all forest management. Thus, it is essentially impossible to segregate forest areas by dominant type of use. Nevertheless, as has already been noted, type of use is significant for pest management policy. Recent trends and present status of each of these major use categories, particularly as it relates to pest management problems, are discussed below.

Soil and Watershed Protection

Almost all forests provide some significant social value through the protection they afford against soil erosion and stream turbidity and through their effect on limiting flood peaks. In areas of steep topography, highly erodible soils, or heavy rainfall, these functions may be crucial to protection of heavily populated areas, as in the Los Angeles and Salt Lake basins. Few natural organisms disrupt this function to a significant degree (although naturally occurring fires may be devastating to this use). Pest damage is usually specific, and even epidemic infestations that virtually wipe out suitability of the forest for other types of use leave sufficient vegetation on the ground to fulfill most of the forest's soil and watershed protective functions. Particularly in the arid West, where over 90 percent of the annual runoff of water originates on forest land, usable water yields may in some circumstances be increased by partial or total removal of the tree cover. In such a case, extensive killing of the trees could increase usable water yields at least for a few years. This theoretically beneficial effect will sometimes be of minimal value in practice, because in some climates, water yield from a forest may be effected chiefly by removing trees from the immediate riparian area, which is a small fraction of the total forest.

The responses of forest managers to a pest outbreak may be considerably more damaging than the pest outbreak itself to soil and watershed values. Improperly controlled applications of pesticides on a watershed could lead to stream pollution. Commercial salvage of pest-killed timber may require construction of roads in previously roadless areas and extension of the logging season into relatively rainy periods. Thus, a pest epidemic may be the indirect cause of soil and watershed damage associated with road construction and off-season logging. (See NAS 1973 for a more detailed discussion.)

Wildlife Habitat

Virtually all forest land provides wildlife habitat, and many species, including the principal big game animals, are found primarily in forested areas. Use of forest land for hunting, fishing, and observation of wildlife has steadily increased. As in the case of soil and watershed protection, most organisms commonly regarded as forest pests appear to have only minor effects on wildlife habitat, because of the forest's ecological diversity. However, a large area kill or defoliation can alter the habitat rather drastically. As in the previous case, pest control measures also may affect wildlife habitat. Improperly controlled pesticide applications, for example, may damage mammal, bird, or fish populations either by poisoning them directly or by changing critical features of the environment in which they live.

Growing and Harvesting Timber Products

Although much of the 750 million acres of forest grows at least some timber, less than two-thirds of this acreage is subject to harvest for timber products. Industrially owned forests, most commercial forest in public ownership, and a significant fraction of other private land either have been logged with some regularity in the past or are included in formal timber management plans. Most forest pest control actions to date have been undertaken in order to protect economic values for timber products use threatened by insects or disease.

The value of timber as a natural resource is very high. It has been estimated that in the United States the weight of harvested roundwood approaches the *total* production of steel, aluminum, cement, and plastics (PAPTE 1973). In addition, standing timber values have risen steadily, increasing some fourfold in real economic terms over the past 70 years.

The rise in these values has both encouraged and been used to justify efforts to extend pest control programs. In the past, most of these pest control measures have been taken to protect mature and overmature forests for which commercial realization value was high; more recently, efforts to protect young stands from mammals, insects, weeds, and pathogens have increased. Forests that are overmature from the point of view of timber growing approach biological senescence, and their susceptibility to certain insects and diseases increases. As a result, epidemic outbreaks of these organisms have often been associated with overmature forests. Organized timber products harvesting has permanently eliminated overmature timber on most commercial forest, with consequent reduction in susceptibility to the pests of overmaturity. That process has changed, and continues to change, the nature of the forest pest problem, but this in no sense reduces its importance.

The forest products industry currently accounts for about 5 percent of the nation's gross national product (GNP) in the form of raw materials for housing, packaging materials, paper, and a multitude of other essential materials. To the extent that forest pests reduce standing timber volume, value, and growth, they contribute directly to the increasing scarcity of the raw material supply on which this sector of the national economy depends.

Outdoor Recreation

Recreational use of forest land includes a great number of quite dissimilar activities. These range from the group viewing the cliffs of the Yosemite Valley from the veranda of a luxury hotel to the mountaineering party in the Brooks Range, and from the motorized family complete with camper, trailbikes, and portable televisions, all installed in the forest campground with hot showers and electricity, to the solitary cross-country hiker.

The amount of forest land used for these varied purposes is even less well known than is the amount used for timber growing, in part because much recreational service is provided by forests that are also used for other purposes. Certain forms of outdoor recreation are either dominant or codominant uses in units of the national and state park systems, national recreation areas, and the wilderness system. Some 60 million acres of land has been formally assigned to these units. In addition, recreation is a dominant use on portions of the national forests outside wilderness areas and is a permitted use on most multiple-use areas of the forests.

The trend in recreational use of forests has been almost explosively upward throughout most of the past 40 years. Except for the period of World War II, attendance rates at national and state parks and national forests have at least doubled during each successive decade.

The very diversity of forest recreation activities makes it difficult to generalize about either the types of pest that affect them or the impact of these pests. The situation in campgrounds, picnic areas, and other sites designed for visitor occupancy may be quite different from the situation in the recreational forest, which simply forms the backdrop for hiking, riding, climbing, or viewing. In the latter sort of area, where recreational use is widely dispersed, the main goal of forest protection is to maintain aesthetic quality. Levels of pest impact which would be considered seriously damaging to a campground or heavily used lakeshore may, in this case, be entirely acceptable. However, severe and widespread insect or disease attacks, such as those of the lodgepole needle miner in Yosemite Park, may create unsightly areas. In such cases, pest control may be thought necessary, even in areas used for dispersed recreational activities.

In campgrounds and other occupancy sites on recreation areas, pest problems are likely to be much more localized but more numerous. At the same time, the relatively high value per acre of such sites may justify quite intensive methods of control. Local elimination of plants and animals poisonous to human beings may be required. Dead and dying trees, in addition to possible unsightliness, increase the hazard to users from both fire and winds--hazards that must be kept at a minimum. Due to the high value, high accessibility, and close surveillance that characterize such areas, intensive and discriminating methods of pest control will usually be feasible.

Forage Production

Forage for domestic livestock is among the by-products of the outputs from forest land. There are, of course, large areas of true grassland within the administrative jurisdiction of such agencies as the USFS. But in addition, several important forest types (e.g., most hardwood types, pine types in the South and West) produce grass and herb ground covers of substantial forage value. Management of these forest types to increase nutritious forage may involve use of selective herbicides, controlled use of fire, or other measures. Unless improperly controlled, such measures are not likely to create substantial pesticide problems.

Preservation of Natural Ecological Features

This use of the forests is of dominant importance in national parks, units of the wilderness system, and some state parks. In general the objective on such areas is to manage them so as to maintain primitive conditions. In addition, a limited acreage has been set aside by various owners in natural areas, primarily for purposes of providing ecological benchmarks and situations suitable for scientific study. Because "pests" are defined solely with respect to human objectives, and because the purpose of preserving these areas is to maintain them as free as possible from human impact, one might expect that forest pest problems would not arise where this sort of use is dominant. In practice, the situation may not be so simple because of the existence of introduced pests that may attack the natural biota, potential spread of pests outside the natural area, and the occasional threat to major aesthetic features.

White pine blister rust, introduced into the country early in this century, is an example of the first problem. This disease can be lethal to all five-needled pines. In some wilderness areas, managers face a true dilemma--either to allow the disease to run its course with consequent permanent change in the plant cover, or to try to protect white pines by means that are inconsistent with the objective of complete noninterference.

Wilderness areas include unusually large proportions of old overmature forest. Such forests are often susceptible to insects and diseases of senescence and may provide conditions favorable to epidemics. If such outbreaks occur, and if the pests are highly mobile, there may be nothing to prevent them from extending their area of infestation outside the wilderness and into forests managed for other purposes, where serious damage may be done.

Production of Christmas Trees, Forest Tree Nursery Stock, and Seed

Both Christmas trees and tree seedlings for forest plantings are mass-produced by intensive cultural methods similar to those of horticulture. Forest tree "seed orchards," where selected genotypes are either grown to seed-bearing size or grafted to established trees, are maintained to provide improved seed for nursery sowing.

All three of these highly specialized tree "crops" are carefully tended throughout their comparatively short rotation periods. Christmas tree plantations are protected from agents causing evident damage to shoots and foliage by methods similar to those used for ornamental plantings. Competing vegetation is now also recognized as a serious obstacle to commercial Christmas tree production, and herbicides are increasingly being used. Forest tree nurseries, with their many young plants in uniform beds, provide conditions that are highly favorable to pest attack, especially by various pathogens. The initial investments that are required to produce forest seedlings, together with the high crop values, provide strong incentives for efforts to control these pests. Management of seed orchards resembles standard orchard practice. Pest control measures are used, especially to reduce damage from cone and seed insects.

In these three special forest areas, the trees or tree products are especially susceptible to particular types of pest problems, either because of the methods of cultivation or because of the nature of the product. All are subject to relatively intensive measures to control those pests, primarily because of the high value of the products.

Suburban and Urban Forest Use

Forests located within urban and suburban areas represent conditions where pest control must be considered in its most intensive form. The values are high and are often assignable on a tree-by-tree basis. Side effects of any control measures are likely to be more critical than elsewhere, and the methods of control may be severely limited because of the close proximity of the human population. Impacts of insects or disease that might well be ignored in other situations because of their ephemeral character (for example, some types of defoliation) may represent significant damage. The problems are often multiplied by the presence of large numbers of exotic species and by cultivation practices such as irrigation and soil manipulation.

Although the circumstances of urban and suburban forestry have, until now, been radically different from those surrounding more conventional forms of forest management, it is becoming increasingly evident that the differences are mainly of degree. As time goes on, and as the intensiveness of forest management increases, the problems of pest control in the commercial and recreational forests will approach comparability with those of suburban forestry, in magnitude of the values at stake, in the need to localize impacts of both pests and protective measures, and in the level of detailed information about relevant ecosystems needed to handle the problems successfully.

INTENSIVENESS OF FOREST MANAGEMENT

Because demands for each of the forest land uses just enumerated have increased over the past several decades, while the area of forest first stabilized and then began a slow decline, the intensiveness of forest management has gradually increased overall.

To fully understand the significance of this trend toward intensified forest management, the meaning of the term must be clarified. Intensiveness of management is often measured by the value of the inputs per acre applied to the land. Thus, intensive forest management for timber production is conceived to mean relatively heavy use of thinnings, fertilization, tree improvement, and the like. But the level of inputs is only one measure of management intensiveness. An equally important measure, and the one to be stressed here, is the minimum size of the land unit used in management planning. In the early days of forestry, the smallest unit recognized for planning purposes was perhaps a mile-square section of land. As values increased and planning funds permitted, the minimum size of planning unit has gradually declined. Today, the minimum size of planning unit in "suburban" forestry may be an acre or two, and on an intensively managed tree farm it may be 10 acres. This dimension of intensive forestry is of tremendous importance because the smaller the basic unit used in planning, the greater the opportunity to recognize the great biological variability of the forest. In the future, pest problems will increasingly have to be considered on the 10-acre planning unit basis, not on the square-mile basis, because of the growing pressures for more intensive use of the land.

REFERENCES

Little, Elbert Jr. (1949) Important forest trees of the U.S. *In* Trees: Yearbook of Agriculture 1949. U.S. Department of Agriculture, Washington, D.C.

National Academy of Sciences (1973) Man, Materials, and Environment. A Report for the National Commission on Materials Policy. Washington, D.C.: National Academy of Sciences.

President's Advisory Panel on Timber and the Environment (PAPTE) (1973) Report. Washington, D.C.: Government Printing Office.

U.S. Forest Service (1973) The Outlook for Timber in the United States. Forest Resource Report 20. Washington, D.C. 367 pp.

Vaux, Henry J. (1973) How much land do we need for timber growing? J. Forestry, 71(7):399-403.

2

THE NATURE OF FOREST PEST PROBLEMS

The structure and function of a forest ecosystem is determined by a dynamic balance between available energy and nutrients, growth, reproduction and natural regulations of its plant and animal components, on the one hand, and factors in the environment that limit or inhibit these processes, on the other. Each component plant and animal population is subject to predation, parasitism, competition, or interference by its own kind or others, as well as to major and minor fluctuations in response to physical forces. Large temporary increases in numbers of one kind of organism at the expense of others and even drastic alteration of all populations and the habitat itself by agents such as drought, wind, and fire are "normal" events when viewed against the life span of individual forest trees, which can range from one to several centuries. Such events, in fact, shaped the structure and composition of the forests found here by European settlers.

This perspective on the forest is important to understanding, but it is neither specific nor useful operationally, except in treating wilderness areas. Man's usual concern with the forest is limited to the more conspicuous species of organisms, to individual trees and stands, and to their development over periods of years or decades rather than centuries. Thus for most purposes, we label uninterrupted (or not visibly disrupted) development as normal, and consider the diverse factors that interfere as "stress." It is convenient to separate the numerous "biotic stresses" from those caused by abiotic physical agents such as weather, fire, or air pollution, although these latter sometimes have a powerful influence on the outcome of biotic stresses.

Because forest trees are long-lived, they are often exposed to a series of biotic and abiotic stresses. At any time during its life span the characteristics and vigor of either a tree or a forest reflect the sum of interactions among all of the growth factors and stresses to which it has been exposed. The pronounced variations in annual ring patterns seen on a cut stump are only one manifestation of these shifting interactions.

The importance of any particular stress factor is relative rather than absolute. The significance of a disease or injury depends both on its interactions and on one's view of the ecosystem or one's objectives in managing a forest. As noted earlier, there is no such thing as a forest pest in an unqualified sense, for it can be defined only in terms of the values man wishes to derive from a particular forest.

The following sections emphasize that pest problems in forestry are sometimes similar, sometimes quite different from pest problems in agriculture. Because of the high value and short rotation times of agricultural crops, and because most agricultural pests have easily identified, direct impacts on annual production, considerable investments have been made in agricultural pest research and pest control. Although forest trees have a relatively high value, the rotation time of the crop is usually very long, and many of the stresses on trees are more subtle and less easy to identify over long periods. These stress factors do not necessarily result in the death of a tree, but they may retard growth and lengthen the rotation time significantly. Thus, the USFS, in its timber inventories, uses the term "growth impact" to describe both loss of growth and mortality of trees. This concept and the ways in which growth impact is measured are now being reexamined by the USFS.

No recent estimates of loss and mortality in the nation's forests are available. In 1952, however, the last year for which such figures are available, the total growth impact on commercial forest lands in the continental United States and coastal Alaska from wildfire, animals, and pathogens was estimated to be over 11 billion board feet (12 board feet = 1 cubic foot), which is the equivalent of 29 percent of the lumber produced in that year. Over 7 billion board feet of growth loss was due to stresses of disease, weather, and animals (USFS 1958). Losses attributable to competing vegetation would increase the total considerably. More recently, the annual loss of timber in Oregon and Washington from disease alone has been estimated to be 403 million board feet, or about 13 percent of annual growth (Childs and Shea 1967). However, in spite of the magnitude of its total impact on growth, pest control in forestry has been applied in relatively local situations where costs have been clearly outweighed by benefits. The reason for this involves primarily the economics of timber growing: benefits from such inputs as pest control are long deferred because of the relatively long rotation time.

The following sections describe the pest characteristics and some general approaches to the control of pathological stresses, insects, vertebrates, and weeds.

PATHOLOGICAL STRESS

The science of pathology deals specifically with those stress factors that cause injury or disease. The various stress factors traditionally considered pathological are numerous and varied, and can be caused by nonliving (abiotic) or living (biotic) agents. The emphasis in this report will be on biotic agents, although abiotic agents will be discussed briefly in terms of their tendency to predispose trees to injury by biotic pests.

Since 1951, the USFS has issued an annual publication entitled *Forest Insect Conditions Report*. In 1971, this publication was expanded to include forest diseases and was renamed *Forest Insect and Disease Conditions in the United States*. This annual publication provides an up-to-date listing of the tree diseases of importance in the various regions of the USFS. A summary of the diseases receiving the greatest attention in the recent reports (USFS 1972b, 1973) is provided in Appendix A of this report.

We have attempted to provide a framework for the understanding of the most important disease agents of forest trees. These include the abiotic stresses caused by adverse soil properties (related to moisture, temperature, aeration, and nutrient availability) and air pollution, and the biotic stresses caused by certain fungi, bacteria, viruses, mycoplasmas, nematodes, and parasitic higher plants (Smith 1970).

Before introducing these various disease agents in more detail, it is important to review several generalizations concerning forest tree disease.

As trees are large and long-lived, their health at any point in time will be determined by the sum and interacting nature of all stresses influencing them. Recent research on forest tree disease has provided much evidence to indicate the importance of interactions between various biotic stress factors, between biotic and abiotic stress factors, and between pathological and nonpathological stresses. For example, injury due to air pollution may predispose trees to bark beetle infestation; insect defoliation may predispose trees to fungal root infection.

Pathological stresses may have considerable ecological significance because of their ability to influence evolutionary adjustments, species composition and diversity, nutrient cycling, succession, and other ecological characteristics. If trees are being grown for a commercial wood product such as lumber or paper, disease and injury may also assume considerable economic importance.

As is true for all pests, however, management objectives are of fundamental significance in determining the relative importance of specific diseases. Microbial decay of stem heartwood, for example, could result in the loss of considerable wood volume in a forest being managed for sawlog or pulpwood production. This decay process would be almost completely without significance if the same forest were being managed for watershed protection or recreation. Root disease that predisposed scattered trees to windthrow would be inconsequential to the manager of a wilderness area but would be of central concern to the manager of a campground.

Relative importance of specific pathological stresses is also determined by location because of variations in climate, proximity to people and their machines and factories, geology, topography, and other factors. For example, the incidence of white pine blister rust caused by *Cronartium ribicola* is lower in southern than in northern Wisconsin because of microclimatic differences (Van Arsdel et al. 1961). Forests in the vicinity of Los Angeles, California; Spokane, Washington; and Copper Hill, Tennessee, have been under air pollution stress because of their proximity to industrial or urban sources. The spread of root rot of Port Orford cedar caused by *Phytophthora lateralis* has been faster in the lowland portion of the range of cedar than in the mountainous portion (Roth et al. 1972). Slash pine is more severely infected by *Cronartium fusiforme*, the cause of fusiform rust, in the northern part of its planted range, while loblolly pine is more severely influenced in the extreme southern portions of its range (Zobel et al. 1971).

The following sections introduce briefly the most important abiotic and biotic tree disease stress factors. More detail on the agents and mechanisms of tree disease can be found in Smith (1970), and comprehensive consideration of specific diseases of forest and shade trees of the United States is presented in Hepting (1971).

Abiotic Disease Agents

The abiotic stresses are briefly mentioned here because they may weaken trees and thus predispose them to disease and to attack by insect pests. Among the most important of such stress factors are adverse soil properties and air pollution.

Tree vigor is critically affected by soil characteristics. Variations in physical and chemical soil properties frequently cause abnormal physiology and result in tree disease. Insufficient soil moisture during periods of drought may cause suppression of transpiration, photosynthesis, nutrient uptake, and growth and may cause death in extreme cases. Excess soil water may alter root respiration and nutrient uptake by creating oxygen stress. Poor aeration of roots can also be caused by fills resulting from grade changes, for example, in areas of roadway construction and in residential subdivisions built in wooded locations.

Soil characteristics often interact with biotic stresses to influence severity of disease. For example, root infection by *Fomes annosus*, an extremely important fungal root pathogen, has been shown to be correlated with soil field capacity, texture, and pore space.

The effects of air pollution on trees are highly varied because of the diverse nature of the contaminants; the influence of local edaphic, topographical, and meteorological conditions; and the considerable range of susceptibility within, as well as among, species. Nevertheless, it has been clearly established that air pollutants injure trees and may predispose them to damage by insects or diseases (Smith 1973, 1974).

Biotic Disease Agents

Fungi

This group contains more complex morphological forms, more diverse life cycles, and more species than any other microbial group, with the possible exception of the algae. There are more than 4,000 genera of fungi and more than 100,000 species. Approximately 8,000 species can cause disease in plants. Fungi are the most important microbial group causing disease of trees. They infect all species of forest trees, at all stages of the life cycle, and in all of the tissues and organs of some species. Following are some of the most important groups of fungal tree diseases:

Wood Discoloration and Decay The phenomenon of root, branch, and stem wood decay by fungi is both an asset and a liability. Because of their virtually unique capacity to degrade lignified tissues, fungi play a critical role in the carbon cycle. As the fungi decay wood, forest debris is diminished, organic and inorganic materials are added to the soil, and carbon is returned to the atmosphere in the form of carbon dioxide. In commercially important trees, however, wood-decay fungi are a liability because these organisms destroy valuable raw materials, especially in mature and overmature trees. In certain commercial forests, wood decay may account for as much as 75 percent of the loss due to disease. For forests or trees managed for purposes other than commercial wood production, fungal decay is largely unimportant because the physiology of the tree is not significantly influenced. Decayed trees are, however, predisposed to windthrow and constitute a hazard to people in urban and suburban locations and in recreation areas.

Rust Diseases Diseases caused by rust fungi are characterized by a wide variety of symptoms. Rust diseases are commonly associated with tree leaves, cones, and stems. The most serious rust diseases of all trees are the stem rusts of conifers. In these rusts, the mycelium of the causal fungus is typically perennial in the bark or wood of the host tree and causes various host malformations, including galls, other tumors, cankers, and witches' brooms. Seedlings and saplings are frequently killed by such infections, while infection of older trees commonly results in disfigurement and reduced vigor.

Blister rusts are extremely important stem rusts on conifers and are caused by fungi of the genus *Cronartium*. Until approximately 1900, the most important *Cronartium* diseases in the United States were eastern gall rust on Virginia and shortleaf pine caused by *Conartium cerebrum*, western gall rust on western hard pines caused by *C. coleosporiodes*, comandra blister rust on lodgepole pine caused by *C. comandrae*, and southern fusiform rust on southern hard pines caused by *C. fusiforme*. In the early 1900s, however, *C. ribicola*, which causes white pine blister rust, was introduced to this country from Europe. The disease has spread through nearly the whole range of our native white pines and has caused destruction of pines resulting in major economic and ecologic impact (see Chapter 4). Because of the intensive culture imposed by man, fusiform rust is presently causing more than $10 million of damage annually in southern forests (Verrall and Czabator 1970).

Canker Diseases Cankers are localized regions of necrosis in the stems or branches of trees. Many fungi cause canker symptoms in both broad-leaved trees and conifers.

Numerous important canker diseases affect conifers. *Atropellis* canker is a widespread disease of pines caused by any of four species of that genus. Malformed stems and discolored, resin-soaked wood make the canker portion of trees undesirable for lumber. *Scleroderris lagerbergii* has caused dieback and cankering of pines and dieback of spruce in Europe and Great Britain for nearly a century. Within the last 20 years, this organism has become important in Canada and the United States. Its greatest impact has been on native red and jack pines, with less effect on the imported Scotch pine.

The significance of canker-type diseases, however, is greater among the hardwoods. Beech bark disease is an important disease of American beech caused by the bark-killing fungus *Nectria coccinea* var. *faginata*, which infects through minute wounds made by the beech scale insect, *Cryptococcus fagi*. This disease, probably introduced from Canada, has spread through the Northeast to Pennsylvania and is spreading southward. Hypoxylon canker is the most important killing disease of aspen. This disease is particularly damaging in the Lake States, where its total impact is estimated to be equivalent to one-third of the net annual growth. The most infamous canker disease of all, however, is the chestnut blight caused by *Endothia parasitica*. This fungus was introduced into the United States on nursery stock imported from the Orient. Spreading at the rate of approximately 24 miles per year, the pathogen rapidly became distributed throughout the natural range of chestnut. Despite all control efforts, *E. parasitica* destroyed every mature chestnut within the natural range in less than 50 years.

Foliar Diseases Numerous fungi cause foliage diseases on both broad-leaved trees and conifers. Generally, the reduction in functional leaf area caused by foliage diseases is less significant in broad-leaved trees. In such trees, even complete defoliation, which is not common, would cause serious damage only if it occurred early in the growing season in several consecutive years. Defoliation does, however, result in reduced resistance to other stress factors and may cause unsightliness in urban and ornamental situations.

Complete or partial defoliation is substantially more important in conifers, whose normal metabolism depends on the activity of several years' accumulation of needles. Any appreciable loss due to disease is thus not followed by a complete replenishment the following growing season. In most conifers, defoliation causes death. Needlecast diseases occur in all groups of conifers--pine, spruce, fir, larch, and cedar. At least forty different species of fungi are known to cause such diseases, and those of six genera are of particular importance: *Rhabdocline* on Douglas fir, and *Lophodermium*, *Hypodermella*, *Hypoderma*, *Elytroderma*, and *Bifusella* on pines or other conifers. Brown-spot needle blight, caused by the fungus *Scirrhia acicola*, is a serious disease of longleaf pine seedlings. The annual growth loss due to this disease has been estimated to be equal to 16 million cubic feet of timber.

Vascular Diseases Fungal infection of the vascular, or water-transport, tissue of trees causes severe impairment of physiological activity, and frequently results in death. These diseases are most important in the angiosperm group, and the general symptom is wilting. Infection of Norway, sugar, red, silver, Japanese, bigleaf and sycamore maples and boxelder by *Verticillium alboatrum* causes maple wilt, an important disease of ornamentals. Oak wilt, caused by *Ceratocystis fagacearum*, is one of the most important fungal diseases of the oak group. Major damage from this disease has occurred in forest stands in Wisconsin, Minnesota, and Iowa. By far the best-known vascular disease in the United States is Dutch elm disease. This malady of all native American and European elms is caused by the introduced fungus *Ceratocystis ulmi*. This disease has had tremendous impact on urban and ornamental elms; the disease is discussed more fully in the section of this report on control efforts in urban and suburban areas.

Root Diseases Root diseases are an extremely important tree problem that affects all stages of the tree life cycle, from seedling to mature tree. Because germinating seeds and young seedlings are succulent and susceptible to microbial infection, many are rapidly rotted by various fungi. This rotting, which may involve virtually any part of the seed or seedling, is called damping-off. Damping-off diseases, which, along with nematode damage, represent the most important diseases of the nursery, injure the seedlings of almost all tree species.

Root diseases also damage older trees. Laminated root rot, caused by the native fungus *Poria weirii*, is by far the most destructive root rot of Douglas fir. This disease causes an estimated annual loss of 32 million cubic feet. Root infection of conifers by *Fomes annosus* also causes very important losses throughout the temperate region. On oak and most other hardwood species, root infection by *Armillaria mellea* is particularly damaging to trees under stress from some other factor.

Bacteria

Bacteria pathogenic to trees cause almost as many symptoms of disease as do fungi. They cause leaf spots and blights, soft rots of fruits and roots, wilts, tumors, scabs, and cankers. In addition to direct disease production, bacteria are thought to be involved in wetwood production and in the processes of wood discoloration and decay. With few exceptions, however, bacterial diseases of forest trees do not appear to have extreme ecologic or economic significance.

Viruses

The significance of viral diseases of several agricultural, ornamental, and tree fruit crops is very well documented. The study of such maladies in forest trees, however, is in its infancy. About 60 viral diseases that affect commercially important forest species have been described (Smith 1970), but this figure may grossly understate the importance of viruses as tree disease agents. This hiatus of knowledge is especially meaningful when one realizes that virus diseases are most serious in perennial plants, since, once infected, such plants remain infected throughout their lives. The manifestation of viral infection may not be acute or dramatic but, rather, subtle and expressed in a gradual decline syndrome. Virus infection is significant in eucalypts, elm, oak, birch, ash, and black locust.

Mycoplasmas

Although mycoplasmas, the smallest free-living microorganisms, have been known to cause disease in humans and animals for a number of years, the first occurrence in plants was reported in 1968. In the past 6 years, approximately 50 diseases, previously thought to be produced by "yellows" viruses, have been proven to be caused by mycoplasmas. Both viruses and mycoplasmas are transmitted from plant to plant primarily by insect vectors.

The significance of disease caused by mycoplasmas in forest or ornamental trees is not known. The following tree diseases may be caused by mycoplasmas: paulownia witches' broom, sandalwood spike, elm phloem necrosis, black locust witches' broom, willow witches' broom, and ash witches' broom. In forest trees, mycoplasma infection has its greatest effect in causing disfigurement, predisposition to wind breakage, and gradual debilitation, rather than immediate mortality.

Nematodes

Nematode damage is well known in nonforest tree crops such as citrus, peaches, walnuts, almonds, and cherries but is of uncertain significance in forest trees, even though nematodes are found in most forest areas. Economic loss due to nematode infection is common in forest tree nurseries. Nursery crops are particularly vulnerable to nematode damage because of the repeated cropping of the same or closely related species on the same area and the abundance of irrigation water and nutrients from fertilizers. The significance of nematodes associated with forest trees in plantations and natural stands is unclear at present (Ruehle 1973). Reduced growth resulting from mass feeding by nematodes appears to be particularly deleterious to marginal trees. The greatest importance of nematodes to tree health, however, may be in their secondary effects. The feeding wounds caused by nematodes may provide entry for root-infecting microbes. In addition, nematodes may transmit viruses from tree to tree. Disruption of mycorrhizal relationships may be an important indirect effect of nematode infection.

Higher Plants

Certain higher plants--for example, true mistletoes, dwarf mistletoes, and dodders--cause significant tree disease in certain situations. Dwarf mistletoes (*Arceuthobium* spp., which are true higher plants with microscopic shoots, leaves, and seeds) are considered the most destructive higher plant disease agents in most western commercial forests. Infection by *Arceuthobium* spp. results in both reduced growth rate and increased mortality. In Oregon and Washington alone, dwarf mistletoes cause an estimated growth loss of 75 million cubic feet annually, along with a mortality of 72 million cubic feet.

Approaches to Control

The control efforts directed toward pathogens of forest trees are not uniform, nor are they necessarily comparable to those employed in agriculture. For a variety of reasons, only forest areas that are quite intensively managed are subject to much artificial control of pathogens. The controls that are implemented depend largely on the economics of the situation and on the biology of host and pathogen.

Most control strategies are cultural, encompassing site and species selection (when establishing a new stand), selection of silvicultural method, determination of rotation length (time to harvest), and operations such as draining, irrigating, thinning, pruning, burning, and fertilizing. The use of pesticides (fungicides, nematocides, insecticides, or herbicides--all of which may control tree disease under some circumstances) has been confined to specialty forests and highly valued individual trees.

On millions of acres of commercial and noncommercial forests, pathological stresses either are not controlled at all or are partially controlled by minor silvicultural manipulations. On the several hundred thousand acres of specialty forests such as nurseries, seed orchards, windbreaks, and Christmas tree plantations, the primary pathological stress controls are both cultural and pesticidal and are frequently intensive. Throughout the approximately 35 million acres of urban and suburban land, too, the primary pathological stress controls are both cultural and chemical and are commonly intensive. Specific control procedures are discussed in the context of management systems in Chapter 3.

INSECTS

Numerous insect species coexist within a forest area with only occasional significance in terms of man's use of the area. Populations of both native and introduced insect species, however, may at times cause considerable economic losses. Perhaps because of spectacular outbreaks, forest insects were of interest for many years even before the development of professional forestry and intensive forest management.

The consequences of cumulative insect damage depend in large part on the situation in which it occurs. As with other pests, the perceived severity of the problem and thus the intensity of control efforts used against an insect are directly related to the value of the trees damaged by the pest. The type of damage sustained is also important. For example, a pest that defoliates trees without killing them may cause growth losses that are not great enough to justify massive control efforts in a lightly managed timber-producing area. In a more intensively managed area, a season's loss of growth may be more significant. In a private recreation area, a season's loss of leaves may mean a season in which the tourists go elsewhere; the owners are likely to do whatever is possible to prevent a year's loss of income, even if the control measures necessary to prevent defoliation are very expensive.

The type of damage caused by insect pests is highly variable. Some insects are injurious primarily to seeds or very small trees, thus interfering with forest regeneration in intensively managed forests. This sort of pest may well increase in importance in the future, as the timber industry comes to depend more and more on plantation forests.

Pests that interfere with tree regeneration have generally been of less concern to forest managers than insects that damage established trees. The latter may be placed in the following broad classes, according to the type of damage they do: defoliators, insects of the cambium and phloem, insects of roots and shoots, and sap-sucking insects. Appendix B lists some of the insect pests that have been important in recent years in various regions of the United States. The major types of forest insect pests are discussed in general terms below (much of what follows is documented more thoroughly in Graham and Knight 1965; and USFS 1972a, 1972b, and 1973b).

Defoliators

Leaf-eating insects, which feed on the foliage of trees, comprise a large group of species. The majority exist in relatively low numbers and never cause appreciable damage, but some species can be extremely damaging. Outbreaks of such insects as the spruce budworm (*Choristoneura fumiferana*) and the Douglas fir tussock moth [*Orgyia* (*Hemerocampa*) *pseudotsugata*] may seem to erupt in a few months' time. In fact, outbreaks usually develop over several years, but damage is often not recognized until the outbreak is near a peak.

The important defoliators include both native and imported pests. For example, the gypsy moth (*Porthetria dispar*) was introduced by man from Europe, and the hardwoods of the Eastern United States happened to provide it with excellent food. The spruce budworm, on the other hand, is a native pest, and has been an important factor in the ecology of the northern spruce-fir forests for centuries.

Defoliation may kill trees, but it often causes only growth losses, as described in the section on defoliating disease agents. The extent of loss is difficult to predict. Many factors are involved, including the relative health of the trees, the habits of the insect, the climatic conditions, and the season. The tree species is also important: hardwoods are relatively resistant and ordinarily will withstand several defoliations, while conifers sometimes die after one complete defoliation. For example, a single complete defoliation of a conifer such as spruce before midsummer is likely to kill the tree. If the tree is defoliated late in the season, however, after buds for the next year have formed, a new crop of needles will develop the following year and the tree will survive, if it does not suffer further defoliation.

Some defoliators, including many of the sawflies, are damaging to young trees. The red-headed pine sawfly (*Neodiprion lecontei*), for example, is a serious pest that often causes mortality in pine plantations. The European pine sawfly (*N. sertifer*) can be very damaging to young Christmas tree plantations of Scotch pine, although the trees seldom die.

Cambium and Phloem Insects

A second type of forest insect feeds primarily on the cambium and phloem of the main trunk and larger branches of trees. This group includes some of the most destructive forest pests. The most important are the true bark beetles, of the family Scolytidae, which excavate egg and larval galleries into the succulent tissues of the inner bark. The bark beetles of the genus *Dendroctonus* are responsible each year for extensive mortality of sawtimber in North America, especially in the South and West. The western pine beetle (*Dendroctonus brevicomis*), the spruce beetle (*D. rufipennis*), the mountain pine beetle

(*D. ponderosae*), and the southern pine beetle (*D. frontalis*) are a few of the primary pests (insects capable of attacking and completing normal development in a healthy, living tree) in the bark beetle group. Other bark beetles attack trees that are already weakened by other forms of damage.

The buprestid beetles of the genus *Agrilus* are another common type of phloem insects. These are found primarily on hardwood trees; they generally attack trees in poor health and often cause mortality.

Of the Lepidoptera in this group, perhaps the most common are the pitch moths which attack pines but generally do not harm dense stands of the trees.

Shoot and Root Insects

Insects that feed on shoots are seldom tree killers, except on small seedlings. Instead, the shoot insects often cause deformities in the trees attacked. For this reason, they may be of great importance when trees are grown for lumber, but less so when pulpwood is the ultimate product. In many areas where white pine is grown, managers have given up planting the species because of the deformities created by the white pine weevil (*Pissodes strobi*). The effect varies with the species of both plant and insect, however, so that generalization is difficult.

The roots of trees are also attacked by insects. Most are enemies of small trees: nurseries are plagued by such pests, and sometimes young plantations are seriously damaged. Well-known root-feeding insects are the white grubs of the beetle family Scarabaeidae and weevils of the genus *Hylobins*.

Sap-Sucking Insects

The sap-sucking insects have modified mouthparts that can remove the sap from living plants without consuming the solid material. The effect of these insects on living trees is much less conspicuous than that of the defoliators, but trees may be seriously injured nevertheless. Sap-sucking insects have been of most concern on orchard and ornamental trees, where their feeding is often intense. Such insects seldom infest large acreages of forest land. An important exception is the introduced balsam woolly aphid (*Adelges piceae*), which has spread fairly rapidly and has caused widespread losses--and may prove totally destructive--of the true firs.

Some of the sap-sucking pests are important in young plantations. The Saratoga spittlebug (*Aphrophora saratogensis*), for example, can kill entire plantations of red or jack pines, but such infestations are generally confined to small areas.

Approaches to Control

Kenneth Graham (1963) has discussed the various aspects of forest insect control in considerable detail. Control should be attempted only after a thorough biological evaluation establishes that damage will otherwise result and after an equally thorough economic evaluation justifies the control effort in terms of the value of the forest and human needs.

Unfortunately, detection and evaluation of insect damage--perhaps the most important component of an insect pest control program--has been given inadequate attention to date. At present, detection work is often done strictly on the basis of visible damage so detection often comes too late for treatment to be most effective.

In some outbreaks, insecticides have been sprayed on a large "buffer zone" surrounding a heavily infested area. This practice may be justified in some, but not all, cases. Buffer zones should be established only in situations where biological evaluation establishes their necessity. The likelihood of spillovers of pests from an untreated area depends on the mobility of the pest and on other specific conditions.

It is often suggested that infested wilderness areas should be treated along with adjoining production forests. Here, too, control should proceed only when the biological evaluation indicates that significant spillover is likely to occur. The overall strategy may also affect such a decision. The current philosophy concerning the control of some defoliators, for example, gives equal weight to the need to save green foliage and to suppressing the pest. With this strategy, a large residual population of the pest may be left in the treated area, and if spillover of pests from untreated areas into the treated area occurs, it is of little significance.

Careful biological and economic evaluations are a necessary part of a sound control program for any forest pest. Specific approaches to control, however, may vary considerably, depending on the type of pest problem.

Principles of control of the four types of insect pests described earlier are discussed below.

Defoliators

Contending with defoliator epidemics poses some of the most vexing and complex questions faced by pest management specialists. The areas affected by outbreaks of a single defoliator species in a single state can be very large--10,000 to 1,000,000 acres or more--and outbreaks are often recurrent or progressive in successive years. The resulting growth losses are certainly enormous, apart from any other consequences, although hard data for large areas are scarce. As an example of those losses, defoliation of water tupelo by the forest tent caterpillar reduces annual growth by about one-half (Abrahamson and Harper 1973), and surveys show that the infested area in southwestern Alabama alone has varied between 11,000 and 800,000 acres per year over the past 12 years. This does not mean that all of the infested area suffered sufficient defoliation to halve growth for the year, but the general magnitude of the loss in this single area is apparent.

Chemicals have been used more commonly against defoliating insects than against other types of forest insect pests. Leaf-feeding insects are exposed and therefore may be readily killed by the less expensive aerial application techniques and with smaller quantities of insecticides. Some of the insecticides used presently for controlling these insects are trichlorfon, methoxychlor, malathion, carbaryl, and Accothion. Other materials, including several microbial insecticides (*Bacillus thuringiensis* and specific viruses) have been used experimentally, with success in some cases. Biological control through the introduction of parasites and predators has been successful against some of these insects; the European spruce sawfly, *Diprion hercyniae*, is an example. Many defoliating insects may be prevented from doing significant damage by the maintenance of habitat conditions that are not favorable to the insect; such preventive methods are not always economically feasible.

Cambium and Phloem Insects

Methods of control of cambium and phloem insects are entirely different from those of defoliators. Aerial treatment with insecticides is generally ineffective. An exception is the use of an aerial spray of methoxychlor to control the smaller European elm bark beetle (*Scolytus multistriatus*), the principal carrier of Dutch elm disease. This is primarily an urban shade-tree problem, and no attempt has been made to treat elms in forest locations.

Lindane, which is applied in solution directly to the bark of the infested trees, is sometimes used in the control of infestations of some of the *Dendroctonus* beetles. Usually the trees are felled before treatment. Another method used against some bark beetle species is the trap tree procedure, in which living trees are felled in an infested area prior to beetle emergence. The newly emerged beetles are attracted to the felled trees and develop their broods there. The infested logs are then removed before the new broods emerge. Many bark beetle species are attracted to freshly felled or windthrown trees and attack those in preference to standing trees. Outbreaks of the spruce beetle commonly begin following wind damage to spruce forests, because the beetles increase rapidly in the windthrown trees.

Birds may be effective in the control of some species of bark beetles. Woodpeckers are often dependent upon the presence of bark beetles in northern forested areas, where other food is not abundantly available during the winter. These birds normally keep populations of beetles at a low level by removing most of the broods from standing trees. Windfelled trees are snow-covered, however, and beetles developing in them may be protected from woodpecker predation.

The major means of dealing with bark beetle outbreaks is salvage of dead or dying trees. Infested trees are removed, and the material is utilized for wood products. Sanitation salvage is a specialized method of removing highly susceptible trees prior to a beetle infestation.

Shoot and Root Insects

The many pests that infest shoots and roots are especially difficult to control. The majority attack young trees; some are especially significant in forest tree nurseries, while others are serious pests in young plantations. Most of them require systemic insecticides for effective control. Mechanical control methods have been used at times, but these require the hand removal of infested tips and can be very expensive.

Preventive means are often employed to avoid problems from these pests. Care in selecting sites is especially important--for example, the European pine shoot moth (*Rhyacionia buoliana*) is seldom a problem on good sites, but where growth is slow, damage from the moth may be serious. Another preventive method is to delay planting a site--for example, a 1-year delay in replanting after harvest may greatly reduce losses due to the pales weevil (*Hylobius pales*).

Sap-Sucking Insects

Control methods for the sucking insects are varied. Some can be controlled by chemicals, but many are very difficult to control because of their tremendous reproductive potential. It has been possible to regulate the activities of some of these insects with the use of biological control methods. Lady beetles, for example, have been used against some of these pests, especially on ornamentals. More detailed knowledge of the biology of the insects may lead to the development of successful biological control schemes for more of these pests in the future.

VERTEBRATE PESTS

Many species of wildlife, as well as domestic livestock (cattle, sheep, and hogs), are found in forests but seldom become pests under most circumstances. Significant damage from wildlife seldom occurs in forests with a diversity of species and populations of vertebrates at densities not in excess of carrying capacity. The damage that is caused is usually insignificant compared with the aesthetic and recreational value of the wildlife. Some timber-harvesting practices, however, cause major habitat alterations that lead to abnormal population levels of a few wildlife species and create conditions that promote local concentrations of potential pests. On public lands that are not intensively managed for timber production, wildlife damage is seldom considered a problem. Most of the high-production forests in the United States, however, are on private lands, where intensive management is practiced and where vertebrate damage may become economically important. This is especially true in even-aged stands in restocked forests that are being managed for especially high yields, such as some private holdings in the Pacific Northwest and the Southeast. (Appendix C lists the important vertebrate pests of U.S. forests, along with their scientific names.)

The impact of vertebrates on timber production can be quite significant. In 1952 (the most recent estimate), the total annual growth impact as a result of damage by animals was estimated at 1,009 million cubic feet (including 2,722 million board feet) on commercial forest land, or about 9 percent of the total impact on growing stock. Only a small percentage of this loss was due to direct tree mortality; about 93 percent of the loss was due to unsatisfactory reproduction and inhibited growth. Most damage from wildlife and livestock occurs during the early regeneration of a forest during which some trees are killed, height development is suppressed, and the length of the commercial rotation is increased. For example, many species of wildlife are associated with early stages of forest development and feed primarily on new growth of herbs, shrubs, and seedling trees. These animals are adapted to habitats that are created by fires, windstorms, and other natural catastrophes, and clear-cut forests and subsequent planting of seedlings or natural regeneration create similar conditions. Wildlife species in this category are deer, elk, rabbits and hares, and various small rodents, all of which can become particularly important pests during forest regeneration.

The extent of damage to timber by livestock varies considerably from one region to the next. In the West and Southwest significant damage is caused by livestock, partly from browsing seedlings but mainly by erosion and site deterioration on overgrazed lands. In the South hogs prevent restocking of longleaf pine; they root up and eat the roots of seedlings and destroy reproduction of both natural and planted stock. Timber damage by livestock browsing is also important in some areas on farm woodlots.

Deer are the principal big game species that become forest pests, although elk and moose are important in a few limited areas in the West. Forest damage by deer occurs principally in the Lake States, the Middle and North Atlantic States, the Rocky Mountains, and the Pacific Northwest. It is ordinarily greatest in the North during the winter months when deer herds are confined by snow to comparatively small areas. However, in some areas with excessively high deer populations, their summer range area has also been affected. Deer sometimes interfere with the establishment of forest reproduction by browsing the terminal shoots and side branches. Less common is the damage caused by big game animals through bark peeling and antler rubbing. Continued heavy browsing can result in deterioration of timber stands through the elimination of the more palatable species and dominance of species that are less palatable. An example is the transition from maple and ash to blue beech, ironwood, and beech in many areas in the Middle Atlantic States.

Damage to timber by bears has been observed mainly in the Olympic Peninsula area of western Washington, in western Oregon, and in California. Bears damage or kill young timber during spring and early summer by stripping bark and eating the succulent cambium layer. Damage tends to be concentrated in small areas and in second-growth stands.

Hare and rabbit damage to commercial forest stands occurs mainly in the Lake States, New England, and in the Pacific Northwest; these animals prevent reproduction by browsing the entire plant in the case of seedlings and by clipping shoots and girdling the main stems of slightly larger trees. Even in more advanced stages of growth, girdling and repeated clipping may retard stand establishment and result in forked stems and bushy trees. Where rabbits are numerous, they are a serious threat to the success of Douglas fir and pine plantations in the Pacific Northwest and in California.

Porcupine damage occurs mainly in the West and the North, and mostly in the winter months when porcupines obtain their food chiefly by feeding on the inner bark and cambium layer of young pines and northern hardwoods. They girdle stems of small trees near the ground, but on larger trees they feed in the upper portion of the bole. The principal damage consists of partial or complete girdling of the trunk and branches. Although few trees are killed outright, growth is reduced and many are deformed or weakened and made susceptible to insect and disease attack.

The killing of trees for food by beaver occurs throughout the range of this animal, but this damage and the flooding of stands are usually limited to small areas and are only locally important in timber management. Pocket gophers and mountain beaver, however, are important pests in conifer plantations, especially in the Northwest, during forest regeneration and have been implicated in widespread plantation failure.

Forest tree seeds are important food for many small mammals and birds, and the impact on establishment of tree reproduction can be severe. The most widespread and important of the seed-eating mammals are the white-footed mice, tree squirrels, and chipmunks. Tree squirrels cause damage not only by consuming seeds but also by cone cutting, often before seed maturity. Tree squirrels are particularly heavy consumers of conifer seeds under natural conditions and have been known to take as much as 70 to 75 percent of the seed crop in some areas.

White-footed mice, because of their fondness for tree seed, their wide distribution, and high reproductive capacity, are sometimes the single most important factor limiting successful forest regeneration, particularly in the Pacific Northwest. Favorable habitat for these mice is created by fire and slash burning and by other methods of clearing land for regeneration. The new vegetation appearing on such areas provides abundant food and results in a buildup of the mouse populations. The increased animal pressure often leaves little opportunity for successful natural or artificial seeding. It has been estimated that a rodent population of 24 deer mice and one or two chipmunks per acre will destroy not less than 80 percent of any conifer seed sown before it can germinate.

Overall, direct damage from wildlife is moderate to locally severe. Certain indirect effects, however, are widespread. Where selective browsing causes failure of the regeneration of desirable species, undesirable brush and weed tree species may become dominant. Thus, animal damage may aggravate the forest weed problems discussed later in this section.

Approaches to Control

There are essentially three approaches to control of damage by vertebrate pests: (1) control of pest populations; (2) protection of trees from the pest; and (3) use of trees that are less susceptible to wildlife damage. Research on all three methods is being conducted by government and industry.

Pest populations, especially rodents, are often controlled by trapping or using poisons. Strychnine has been the most used poison, but sodium fluoroacetate (1080), endrin, thallous sulfate, and zinc phosphide are also employed. A recent development involves the use of foam as a carrier for OMPA (Schradan), a highly toxic but short-lived organophosphate, for control of mountain beaver in the Pacific Northwest. The foam is sprayed into or at the mouths of mountain beaver burrows where it is picked up by the animal's feet.

Poisons for controlling vertebrates are usually placed by hand at bait stations or in burrows, but recently there has been increasing use of poisons to impregnate or coat seeds that are distributed aerially. Endrin is commonly used to protect conifer seeds against rodents, as a 1 to 2 percent concentration applied on ½ to 1½ pounds of conifer seed, or about 2 to 5 grams active toxicant per acre. Wheat treated with 1080 is also distributed aerially to control rodents, but the danger of secondary poisoning of other wildlife has led to a reduction of this practice.

Game species such as deer, elk, and bear are protected by law and can be controlled only by regulated hunting during established game seasons. This form of control is not particularly effective, and, in any event, there is considerable public pressure against large reductions of deer or elk populations, regardless of recommendations of wildlife managers.

Ways of protecting seedlings during forest regeneration include physical barriers, ranging from plastic tubes around the seedlings to fences around entire plantations, and chemical repellants. Tetramethyl thiuram disulfide (TMTD) and Zinc dimethyldithiocarbamate-cyclohexamine (ZIP) are widely used on nursery stock and ZIP is also used in the field for protection against deer. Repellants offer some protection between the time of planting and the initiation of new growth, but they have to be applied repeatedly for complete protection.

Douglas fir and ponderosa pine have been the principal species used in reforestation in western forests, but true firs, hemlock, and spruce are less attractive to wildlife that browse on conifers and are, therefore, less susceptible to damage. The growing market value of the latter species, which have become important pulpwoods, may result in their wide use in reforestation trees, thus reducing wildlife damage. This trend is largely governed by the market and is not due to direct manipulation of forest habitats to reduce the pest problems. Reforestation with large seedlings has been shown to reduce wildlife damage significantly; however, there is a trend toward increased use of small, containerized seedlings which are more susceptible to damage.

In summary, most damage from vertebrate pests in forests is caused by only a few species, the most important of which are deer, pocket gophers, porcupines, and rabbits; and most wildlife damage occurs during reforestation. There are relatively few data on the economic importance of damage, except for a few regions, or on the amounts of pesticides being used or costs of control. It would appear, however, that very little is currently being spent on control of vertebrate pests in forests, except on a few privately owned forests that are under intensive, high-yield management.

PLANT PESTS

In this section, we discuss various vegetation types that are sometimes "pests" in relation to management objectives. Central to this subject is the concept that successful achievement of management objectives by itself has a major environmental impact. This must be kept distinct from the impacts of the specific control practices used to achieve them. Thus, the establishment of a Douglas fir forest or a stable shrub community has an effect that may last for centuries, encompassing all life systems, regardless of the method used to establish such a community.

The classification of a plant as a "weed" pest, perhaps more than that of any other type of forest pest, depends on the situation in which the plant is found. Control of species composition in forests is generally a matter of selective vegetation removal toward some particular end. A number of different objectives are possible, in some cases contradictory ones. A general treatment of plants as management problems in forests is given in Newton (1967).

Management objectives determine whether a plant is a weed on a particular site. Commodity-dominated management depends on replacement of stable brush or noncommercial trees with valuable species that may also be stable; the same stability of brushfields is regarded as a virtue in preventing the establishment of trees on rights-of-way. Although the objectives of management differ, the concepts of vegetation management are common to a wide variety of objectives.

It is an ecological truism that a site will be reoccupied by vegetation after a disturbance; and the management of vegetation resources relies entirely on disturbances of various sorts. Dominant trees and shrubs left after partial clearing of forests are especially vigorous in utilizing site resources and open space (Newton 1973). These may be either desirable or undesirable species. Thus, community stability can work either for or against management objectives. If the species are undesirable, then the more stable the system that becomes established, the more skill is necessary to effect change with a minimum of undesirable side effects.

Plant Pests in Production Forests

In special-use situations, such as seed orchards and tree nurseries, weeds are regarded as critically limiting on production. On the much larger acreage of commercial forest land where trees are grown for timber, however, weed species are only beginning to receive attention commensurate with their impact on productivity.

In production forests, weeds may include trees of the same species as the crop tree or of different species, various shrubs, or herbaceous cover. The unwanted plants interfere either by preventing the regeneration of the desired species or by competing for site resources after a stand is already established.

Weeds That Prevent Regeneration

Prompt replanting of cutover lands is an effective and widely used means of ensuring regeneration in the presence of weed species. Where the land is cleared without reforestation, whether because of wildfires or lack of funds or interest on the part of the landowner, shrubs and grasses often increase in coverage or invade fairly rapidly. Sometimes acreage that has never been managed constructively will be overgrown with noncommercial species. Subsequent management for timber production will require removing or controlling the unwanted vegetation by mechanical or chemical means. The degree of control required varies with the species to be planted.

Weeds That Compete for Site Resources

Low-grade tree or shrub species compete with more valuable species to some extent on nearly all of our forested lands. The reasons for weed dominance vary from place to place. Many stocking problems are a result of man's activities. These include logging without reforestation; disruption through mining, railroad, and grazing activities; and selected harvesting of high-grade trees. The last practice, which causes a gradual deterioration in quality, if not quantity, of production, has been of particular importance. Over a span of up to 300 years, man has continually removed from forests the trees that he finds most valuable. This has left an increasing proportion of trees of submarginal value, including those not well adapted to manufacturing, those that are too small to be usable, and those with insect- or disease-caused defects. A low-value forest remains in many areas.

Not all weed problems are man-made. Extensive fires have allowed the invasion of brush in some areas. In other areas, the natural vegetational trend results in tree species that are less desirable from the land manager's point of view. In parts of the South, for example, pine species may be replaced by a variety of hardwood species if the successional trends are left unchecked. Where the forest is managed for maximum timber production, the softwood species are often more desirable because of their faster growth rates and because there is a ready market for them. Management for softwoods in those areas requires periodic destruction of invading hardwoods, or management to minimize their intrusion. These practices may substitute for wildfires, on which natural pine stands usually depend, but which are unacceptable in present-day management.

The impact of undesirable vegetation on lands managed for timber production is undoubtedly substantial, although it is difficult to measure. Walker (1973) estimated the total acreage of commercial forest land supporting important amounts of undesirable vegetation at some 300 million acres. The trees on this very large acreage all suffer some loss of potential annual growth increment, a loss that may be as high as 55 percent. (This figure is based on an estimated average productivity of 25 percent, with 80 percent of potential assumed to be a realistic production goal in native species.)

Despite this substantial growth impact, the concept of forest weeds is not often given much attention. Research on growth losses is focused on insects and diseases. The USFS has conducted studies on cull tree growth, and several regions have brush or weed-tree control projects. Overall, however, research on weed impacts and controls is very meager in relation to the scope of the problem, and little thought has been given deliberately to needed research on weed species in forested areas.

One reason for this is the nature of this loss. Losses due to weeds are losses of potential future gains, rather than the visible loss of growth that has already occurred. Also, weeds, unlike insect and disease pests, exert their influence over long periods of time. Thus, the weeds that are causing losses now, if not removed, may still be causing losses 20 years from now.

Overstocking, where too many trees of desirable species tie up site resources in trees that will never be utilizable, is a special type of weed problem. Experiments with precommercial thinning demonstrate that reduction of stocking can improve the productivity of much overstocked land (Berg 1970). This is especially true of the Pacific Northwest region, where unit productivity is the highest in the nation.

Plant Pests of Nontimber Forest Areas

Rangelands

About 630 million acres of rangelands are grazed in the United States, much of it in the Rocky Mountain region. On some of this acreage, forest management for timber competes with management for grazing and (on public land) for wildlife forage. In general, however, the grazing resource has been considered of primary importance. The scattered trees, then, along with a diversity of shrubs, are the "weeds," and grasses and forbs are the "crop." Because these areas, properly managed, are not forested areas, they will not be considered in detail here.

Recreation Areas

Where land is managed primarily for recreation, no specific plant is undesirable in its own right, except perhaps one that is poisonous to man. Vegetation in such areas is sometimes manipulated to provide a better wildlife habitat; such treatments may or may not act in direct opposition to the production of timber. Where a recreational facility is heavily used, it is sometimes necessary to clear out dead or dying trees that create hazards of fire or windfall. Sometimes relatively mature stands of timber are cut in order to provide ski slopes. In general, however, it has been unnecessary and economically impractical to attempt to change the species composition of a forested recreation area.

In the West, many of the lands designated as recreation areas have never been deforested, and change would be unnecessary even if the land were managed for timber production. In the populous areas of the East, however, much of the forest land was once cleared for agriculture and is now dominated by what would be weed trees in a production forest. Today, a large part of that wooded land is valued primarily for recreation or residential use, so commercial forest productivity is not its major value. In view of the uncertainty of future population shifts, the status of much of this land may change; exclusion from the future resource base may be unwarranted.

Rights-of-Way

It has been estimated that nationwide there are well over 60,000,000 acres of rights-of-way (Egler 1954). These vegetation-covered areas include roadsides, electric and telephone utility rights-of-way, and those maintained by railroad and natural gas companies. Firebreaks represent additional acreage, concentrated in the Southwest.

While not all of this acreage is within forest land, a large portion is located in forested or formerly forested areas. In these areas, the primary management objective is the removal or prevention of trees that would eventually grow or fall into overhead lines or interfere with sight lines along roadsides.

Approaches to Control

Plant pests, in general, have a regional nature. Western areas, with dry summers, can stabilize in either grass or shrubs. Humid regions may stabilize in shrubs, but generally not in herbs. In the absence of such a shrub cover, a stand of trees, especially shade-tolerant species, can form a relatively stable vegetation type in both areas. These concepts are helpful in reaching management objectives for either rights-of-way or timber management.

Nearly all vegetation control in forests has been carried out either in intensively managed production forests or along rights-of-way. The control methods used on these two land areas are tactically similar, although the objectives, as described above, are quite different. Control of plant pests, unlike control of other pests, is only one part of an overall attempt to promote the development of a certain type of stand. Thus, in commercial forestry, the objective is the promotion of a stable forest of valuable trees; along a right-of-way, the objective is to create a stable cover of shrubs or herbs. In both cases, unwanted vegetation must be controlled to release site resources for the establishment or enhancement of the desired cover. Following are some of the methods used in plant pest control.

Site Preparation

Mechanical scarification is widely used to prepare sites for planting. This approach involves the physical destruction of all vegetation, and tends to cause maximum soil disturbance. It is highly successful in eliminating woody cover but is also characterized by large impacts on the ecosystem.

Broadcast application of herbicides from aircraft has gained in importance. This method generally involves the use of phenoxy herbicides, especially 2,4,5-T, which provide fairly selective control of deciduous plants with minimal injury to conifers. Single aerial applications of 2,4,5-T are less effective in killing vegetation than even moderately intensive mechanical preparation, but the chemical method is also less costly and has no physical impact. Herbicides leave the soil intact and fail to reach or to damage seriously most ground cover under the brush canopy. Animal habitat generally sustains minimal disruption. Thus, even repeated application of short-lived herbicides or such herbicides combined with minimal mechanical treatment create less drastic surface disturbance than full-scale mechanical preparation.

Burning is another effective means of removing unwanted vegetation for site preparation. In the South, many existing stands have arisen as a consequence of repeated fire without any additional treatment. Fire is less disturbing to the soil than is scarification, but there are problems of timing, estimation of fuel moisture control, and smoke management. The reduction of cover sometimes offers temporary relief from animals that damage regenerating trees.

Even after good site preparation, many planted areas need treatment in their second year, and some require further treatment after several years. Selective herbicides can be used on conifers. The removal of weeds from hardwood plantings, where practiced, is still largely mechanical or manual.

Removal of Competing Tree Species

The enormous volume of cull tree material that could be harvested in weed control operations in mature forests must be considered a resource. Most of it can be used as pulpwood with present-day technology, and other end uses are under development. Its present low value is due to the availability of other sources of higher-quality wood fiber with lower labor requirements.

In the absence of increased demand for low-value wood, which would allow economic harvest of thinned material, forest weed control involves several chemical and non-chemical methods of killing unwanted trees. The nonchemical methods include girdling, felling, and the use of prescribed fire; the chemical means include tree injection, basal spraying, and aerial spraying. All are more or less effective means of diverting resources to high-quality trees, but each has a different environmental impact.

Girdling is high in cost and not always effective, but its environmental impact is slight and temporary. Very few trees are girdled today. Injection of individual trees is more effective and lower in cost and may be accomplished with nonpersistent herbicides. Several hundred thousand acres, primarily in the South, are treated in this way each year. Basal spraying is also effective but may be more costly; its impact is similar to that of injection. Its use is primarily confined to rights-of-way. Felling is costly and has a substantial physical impact, and sometimes increases fire hazard by providing fuel. Felling is used for most thinning and right-of-way preparation today. Aerial spraying with rapidly degrading herbicides may be used effectively on shrubs, but it is not successful for large cull trees. Aerial sprays of currently registered herbicides have a general effect on ecosystem structure. As most of the herbicides are of short life and low toxicity, effects on wildlife are primarily related to habitat change. Prescribed fire may be used at frequent intervals to prevent or remove the understory of shrubs and hardwood trees that commonly develops in pine stands. This treatment is effective under some conditions, especially in the South, and it is also finding some use in the Pacific Northwest, where herbicides are used to prepare fuel by desiccation.

The forest weed problem is more likely to require man's intervention than other pest problems. Once dominant desirable trees are established, however, they tend to remain dominant, so that continued trouble with weeds is unlikely. Therefore, a management plan that includes the harvesting or killing of low-value trees and provides for the establishment and culture of valuable species, eliminates the weed problem (Newton 1973).

Precommercial Thinning

The methods used in precommercial thinning of overstocked stands are similar to those discussed above, except that, because the same species are removed as are being grown, selectivity must be by manual means. Most of the work is presently done by chain saw. The use of power saws to fell unwanted trees has some inherent problems: saws have a high injury rate among laborers; cut slash is a fire hazard; insects, especially *Ips* spp., may multiply in slash; and diseases, especially root rots such as *Armillaria* and *Fomes annosus*, can enter fresh stump wounds and from there move into living root systems. The use of chemicals involves a different set of problems largely concerning crew training and discipline. Although these have proven very difficult to overcome, there has been a recent increase in thinning operations using injection of low-toxicity organic arsenical herbicides. The chemical method involves less physical impact than manual thinning, and provides protection against insects and diseases.

In the Pacific Northwest, a combined total of about 100,000 acres per year are given precommercial thinning. The USFS thinned a comparable acreage in that region in 1973. The costs are high at $30 to $100 per acre, but the practice is gaining acceptance.

Right-of-Way Management for Vegetation Control

Herbicides have been widely used in right-of-way management for vegetation control in this country. In 1969, almost half of the 2,4,5-T used in the United States was applied to over 2 million acres of right-of-way (USDA 1971) (this figure does not include rights-of-way treated by federal agencies). Other herbicides as well as 2,4-D have also been widely used, almost always as blanket sprays.

The use of blanket sprays, with heavy dependence on the phenoxy herbicides, is often ineffective in terms of the ultimate objective, which is essentially the control of tree growth. With the most widely used phenoxy herbicide, a 2,4-D/2,4,5-T mixture, a grassland is often the resulting cover after repeated applications. In most forested regions with moist summers, this vegetation type is readily invaded by tree seedlings from the contiguous forest, especially if the grassy cover is discontinuous. The result, then, is a cover type that tends to perpetuate the problem that one is attempting to solve.

A technique that would result in the least disturbance to the existing vegetation and in the process create a shrub cover that would tend to arrest tree reproduction would be preferable. This is the opposite of brush control in forest plantings. In forest management it is well documented that dense shrub covers often necessitate the use of herbicides in order to open site conditions for forest regeneration. It is desirable to avoid tree regeneration along rights-of-way.

Two decades ago Egler (1954, 1958) advocated the selective use of herbicides in right-of-way management, in which undesirable tree growth is removed and the low ground cover, especially shrubs, is preserved. Although this approach has been employed, it has not been widely accepted in commercial application. Inadequately trained maintenance personnel and failure of the industry to advocate the selective approach have tended to perpetuate the use of brush control techniques that do not discriminate among the various types of vegetation.

Although it has been argued that selective techniques are less economical than broadcast sprays, the relative economy depends upon whether one's point of view is short-term or long-term. Single blanket spray applications may be less costly than selective sprays, but repeated blanket spraying is required to obtain adequate control. If unwanted trees are root-killed by selective stump or basal techniques and a plant cover is created that tends to inhibit further tree establishment, the need for future spraying is minimized. At the Connecticut Arboretum, there is a right-of-way demonstration area in which a relatively stable mosaic of shrub communities has been created (Niering and Goodwin 1974). It meets the needs of the utility company and provides high stability and wildlife, floral, and open space values. The selective approach was employed to create this pattern.

The fact that certain utilities, such as those in Connecticut, have essentially converted to the selective approach indicates that it is commercially feasible (Crain 1969). Public pressure has played a role in changing vegetation manipulation practices (Goodwin and Niering 1959, 1962; Niering and Goodwin 1974) and will continue to be important in the future as citizens become more aware of the value of the right-of-way resource.

Forest Roads and Firebreaks

The maintenance principles discussed for right-of-way management are also applicable to vegetation management along forest roads within state, national, and commercial forests. The two major management tools that have been used to maintain sight lines along forest roads are mechanical cutting and herbicides. The use of broadcast foliar sprays may destroy desirable herbaceous cover, and by drift may also affect adjacent nontarget vegetation.

In lieu of this practice, a selective approach should be employed along permanent forest roads. This would involve the basal or stump treatment of only the undesirable growth, primarily trees, that interfere with sight lines. This is the approach recommended and adopted for roadside and utility right-of-way management in sections of our national forests (McQuilkin and Strickenberg 1951, USDA 1966). Along temporary forest roads or where the selective approach is not applicable, the cutting techniques employed in certain commercial forests in the Pacific Northwest appear to be most effective.

Along firebreaks, cutting or selective spray techniques can also be used. Blanket sprays often result in a grassy cover that accentuates the fire hazard. An open mixture of broad-leaved herbaceous plants and low-growing shrubs may constitute a much more desirable plant cover in certain regions. The regional vegetational pattern will tend to dictate the most appropriate techniques.

REFERENCES

Abrahamson, L. P., and J. D. Harper (1973) Microbial insecticides control forest tent caterpillar in southwestern Alabama. USFS Res. Note SO 157. p. 3.

Berg, A. B. (1970) Precommercial thinning in Douglas-fir. *In* Management of Young Growth Douglas-Fir and Western Hemlock. Corvallis: Oregon State University School of Forestry.

Childs, T. W., and K. R. Shea (1967) Annual losses from disease in Pacific Northwest forests. Pac. N.W. For. Range Exp. Sta. Res. Bull. PNW-20.

Crain, H. J. (1969) Preservation of desirable species through selective use of herbicides. Northeast Weed Control Conf. Proc. 23:294-297.

Egler, F. E. (1954) Vegetation management for rights-of-way and roadsides. Smithson. Inst. Rep. 1953:299-322.

Egler, F. E. (1958) Science industry and the abuse of rights-of-way. Science 127:573-580.
Goodwin, R. H., and W. A. Niering (1959) The management of roadside vegetation by selective herbicide techniques. Conn. Arbor. Bull. 111:4-10.
Goodwin, R. H., and W. A. Niering (1962) What is happening along Connecticut's roadsides. Conn. Arbor. Bull. 13:13-19.
Graham, K. (1963) Concepts of Forest Entomology. New York: Reinhold Publishing.
Graham, S. A., and F. B. Knight (1965) Principles of Forest Entomology. New York: McGraw-Hill. 4th ed.
Hepting, G. H. (1971) Diseases of Forest and Shade Trees of the United States. Agriculture Handbook No. 386. Washington, D.C.: U.S. Department of Agriculture.
McQuilkin, W. E., and L. R. Strickenburg (1951) Brush Control in Practice in the National Forests. Northeast For. Exp. Sta. Res. Paper No. 148.
Newton, M. (ed.) (1967) Herbicides and Vegetation Management in Forests, Ranges and Noncrop Lands. Corvallis: Oregon State University School of Forestry.
Newton, M. (1973) Forest rehabilitation in North America: Some simplifications. J. For. 71(3):159-162.
Niering, W. A., and R. H. Goodwin (1974) Creation of relatively stable shrublands with herbicides: Arresting "succession" on rights-of-way and pastureland. Ecology 55:784-795.
Roth, L. E., H. H. Bynum, and E. E. Nelson (1972) Phytophthora root rot of Port-Orford-cedar. Forest Pest Leaflet 131. Washington, D.C.: U.S. Forest Service.
Ruehle, J. L. (1973) Nematodes and forest trees--Types of damage to tree roots. Ann. Rev. Phytopathol. 11:99-118.
Smith, W. H. (1970) Tree Pathology: A Short Introduction. Academic Press, N.Y., 390 pp.
Smith, W. H. (1973) Air pollution and urban woody vegetation. Proc. Urban Forestry Conf. Univ. Wisconsin, Stevens Point, Wis. March 21-22, 1973.
Smith, W. H. (1974) Air pollution--Effects on the structure and function of the temperate forest ecosystem. Environ. Pollut. 6:111-129.
U.S. Department of Agriculture (1966) Vegetation Management for Rights-of-Way; Selective Maintenance for Improved Wildlife Habitat and Scenic Values. Washington, D.C.: Forest Service, Eastern Region.
U.S. Department of Agriculture (1971) Restricting the Use of 2,4,5-T: Costs to Domestic Users. Agricultural

Economic Report No. 199. Washington, D.C.: Economic Research Service and Agricultural Research Service.

U.S. Forest Service (1958) Timber Resources for America's Future. Forest Research Report No. 14. Washington, D.C.

U.S. Forest Service (1972a) USDA Final Environmental Statement on Cooperative 1972 Gypsy Moth Suppression and Regulatory Program. Washington, D.C.: U.S. Department of Agriculture.

U.S. Forest Service (1972b) Forest Insect and Disease Conditions in the United States, 1971. Washington, D.C.: U.S. Forest Service.

U.S. Forest Service (1973) Forest Insect and Disease Conditions in the United States, 1972. Washington, D.C.: U.S. Forest Service.

Van Arsdel, E. P., A. J. Riker, T. F. Kouba, V. E. Suomi, and R. A. Bryson (1961) The Climatic Distribution of Blister Rust on White Pine in Wisconsin. Lake States For. Exp. Sta. Paper No. 87; pp. 1-34.

Verrall, A. F., and F. J. Czabator (1970) Fusiform Rust of Southern Pines. Forest Pest Leaflet No. 26. Washington, D.C.: U.S. Forest Service.

Walker, C. M. (1973) Rehabilitation of Forest Land. J. For. 71(3):136-137.

Zobel, B., R. Blair, and M. Zoerb (1971) Using research data--Disease resistance. J. For. 69:486-489.

3

CONTEMPORARY PEST CONTROL PRACTICES

OVERVIEW

We have pointed out above that what one defines and identifies as a "pest" depends on the objectives prescribed by the land manager. A significant fraction of the forest land in the United States is owned by persons whose objectives do not include any sort of resource management goal. For example, some land is held solely as speculation: the owner anticipates a profit through future sale, but any intervening use of the land is essentially irrelevant. Other land is owned by persons who, because of their place of residence or other circumstances, are so detached from the land that they simply have no occasion to formulate management objectives. Unless outside agencies intervene, neither pest problems nor pest control activities exist on such land. Absence of pest control problems in such areas thus derives from the nature of these kinds of ownership, not from explicit and careful consideration of a land management problem.

Elsewhere, the situation is very different. On most public and industry lands and on portions of the nonindustrial private forest, ownership objectives are rather clearly defined, and the problems of pests and their control emerge clearly. Within this sector, the nature, extent, and strategy of pest control programs will vary greatly, depending on the intensiveness of management of the forest as well as on the identity of the pest and the purposes of management. The general relationship between intensiveness of management and the degree of reliance on human intervention in control programs is indicated schematically in Figure 3-1.

As one moves from left to right on the horizontal axis of the figure, the intensiveness of management increases. The examples of forest situations that fit at various

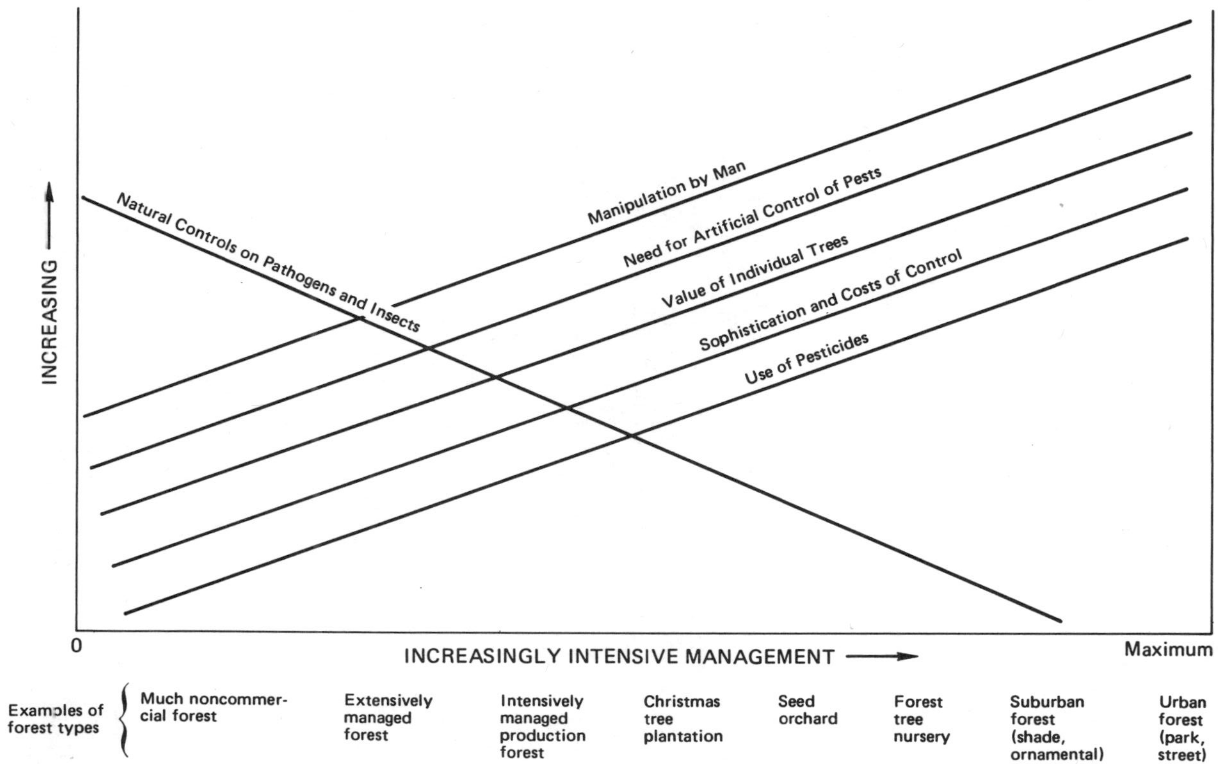

FIGURE 3-1 Characteristics of various forest management situations in regard to pest control practices.

places on the scale can only be ranked very approximately and with considerable overlap. Nevertheless, they are generally characterized by the following (as shown in Figure 3-1): decreased natural controls of native insect and disease pests; increased manipulation by man; increased need for artificial control of forest pests; increased value of individual trees; increased sophistication and cost of control efforts; and increased use of pesticides.

It should be emphasized that the area of forest land to the left of the "intensively managed production forest" category is much greater than the area of all of the land to the right of and including that category. We have found it impossible to assign acreage figures to all the categories because land-use data are not compiled on this basis. Nevertheless, it is clear that the area untreated, or treated only rarely or with extensive cultural practices, is far larger than the area that is intensively treated.

The control measures that are used in these different situations vary considerably. In the least intensive situations, including much noncommercial forest, extensively managed forest, and even a part of intensively managed forest land, low-order infestations and losses from a great variety of agents are simply accepted and tolerated. This strategy is not necessarily passive. Unless the value of the trees is quite high, or the expected damage quite severe, it may well be uneconomic to attempt artificial control of forest pests. At times, too, this acceptance constitutes recognition that "natural control," the action of naturally occurring parasites and predators, is maintaining the pest population within certain limits. Acceptance of injury, then, is a strategy quite different from the absence of pest control on lands where management objectives have not been formulated. Acceptance of injury is the single most widely used strategy on forest lands today.

Legal Procedures

Legal procedures, such as state and federal seed certification programs and plant quarantines, are governmental mechanisms for reducing forest pest problems. Plant quarantines limit the opportunities for new pests to enter the country and often retard their movement on soil, nursery stock, Christmas trees, and logs within the country. The number of serious pests that preceded or evaded quarantine regulations--such as those that cause white pine blister rust, chestnut blight, and Dutch elm disease, and the gypsy moth and European pine shoot moth--suggest the urgent need to prevent further importation. Seed certification requirements reduce unwitting use of poorly adapted or pest-prone varieties and control origin of species planted in forest and parklands.

Other laws and regulations are aimed at diminishing pest problems, directly or indirectly. Legal restrictions against growing black currants and gooseberries, the alternate hosts of the white pine blister rust, still exist, although they are now seldom enforced. Comprehensive forest practice laws that require prompt regeneration in effect prevent leaving cutover lands open to occupancy by woody weeds.

These legal procedures may require the active participation of the land manager, but they differ from other control methods in that they are generally not a part of his operational strategy and discretionary response to damage or threat of damage. They are, however, a part of the background to the manager's decision-making process, and apply, with varying degrees of importance, to all forest situations. In extensively managed and noncommercial forests, these laws may constitute the only defense against pests.

Cultural Control

"Cultural" control methods, including all manipulations of stand growth designed to maintain the overall health and vigor of the trees, are one step up in intensiveness from acceptance of pest damage or management only as the law requires. Cultural methods may include strategies of "avoidance." For example, where trees are planted by man (in plantations or in urban areas), those trees that are highly susceptible to diseases and insects known to occur in the area can be discriminated against, either by choice of species or by genetic selection of resistant stock. The techniques of genetic selection for resistance, and the progress that has been made, are discussed in the section on intensively managed production forests. For certain insects and diseases, it is possible to establish geographic or climatic "hazard zones" that indicate degree of risk (Wilson and Kennedy 1970). Stand vigor may be increased by thinning or harvesting of high-risk species and individuals. Similarly, stocking on drought-susceptible soils may be reduced to improve resistance to associated effects when future droughts occur.

Many weed-control operations are in some ways avoidance activities, as well. For example, plants that might interfere with the regeneration of the desired species are suppressed at the beginning of the life of a stand; and precommercial and commercial thinning operations are carried out to reduce or avoid increasing competition within a stand.

From another viewpoint, many of what are now seen as serious pest problems have been inadvertently provoked in some measure by man's activities. These include poorly stocked or brushy lands resulting from wildfire or lack of prompt regeneration; dominance of uneconomic or cull trees caused by selective removal in the past; excessive deer herds generated by disappearance of predators, protective laws, and altered habitats; and various diseases accentuated by air pollution, compacted soils, or stem wounds. Correction of these effects where possible, or avoidance for the future, will decrease damage later.

Another cultural technique, sanitation, implies positive measures in response to specific threats. It consists of those methods and modifications of operations undertaken to deny particular pest populations the opportunity to persist, to increase their numbers, or to infest. Its role is sometimes seen as passive, as it seeks to prevent, but it is better viewed in the context of sound ecosystem management. In any case, the technique may be totally ineffective against aggressive or virulent pest species, high populations, or distant sources of infection. A common application of sanitation is the removal of overmature, damaged, or otherwise susceptible trees in which the population of certain bark beetles may begin a rise to epidemic proportions. At other times sanitation may be combined with salvage or thinning or may even merge into actual control, as when American elms infected with the fungus causing Dutch elm disease are cut down and destroyed in order to protect remaining elms, or when prescribed fire is used in seedling longleaf stands to destroy the infested needles in which the brown spot fungus sporulates. Other examples are removing alternate hosts of rust fungi, reducing preferred species fed upon by certain defoliators, poisoning prolifically seeding weed trees, such as alder, to reduce encroachment into adjacent openings, and facilitating hunting pressure where excessive deer herds inhibit regeneration.

Sanitation cutting is often combined with salvage, but the latter generally involves the removal of valuable trees that have already been killed or seriously damaged. Prompt action is often necessary to harvest utilizable wood before the dead or damaged trees begin to deteriorate, especially in warm weather, when deterioration occurs more rapidly. Most hardwoods lose much or all of their potential for use as high-value lumber unless harvested before death, and hardwood logs can rarely be stored through the summer months. Small-diameter pine rapidly loses its value for quality sawtimber, although it can be utilized for pulp despite some deterioration of the sapwood. Pine can be stored wherever ponding or spray facilities exist to keep the logs moist.

The possibilities for economic salvage are greatly affected by accessibility and markets. Even where roads exist, large epidemics sometimes saturate the local capacity for logging and utilization or greatly depress prices. However, salvage cutting can significantly reduce the net losses of income to the forest owner and of resources to society after an outbreak of destructive insects or disease.

All of the above cultural techniques are characteristic of the intensively managed production forest and are a part of some control situations in all forest categories to the right of that category in Figure 3-1. Some may be applied in unusual cases even in extensively managed forests.

Direct Control

Other techniques act more directly than those discussed above and involve the active suppression of pests. These include the use of chemical pesticides, prescribed fire, and physical methods.

Fire and physical means are used primarily in the control of weed pests. Both herbicide applications and mechanical site-preparation methods are used routinely in regenerating forests and in preparing nonstocked lands. Mechanical methods are still the most common means of preparing sites. However, according to a recent study (USDA 1972), nearly 500,000 acres of forest plantings were reported by the states as treated with herbicides in 1968, and this figure may be low. Nearly all of the states surveyed thought that the trend in herbicide use for site preparation would be upward, and there are indications that the acreage treated with herbicides has in fact increased rapidly in recent years. A Weyerhaeuser spokesman reported that site preparation on that company's land was about half mechanical and half chemical by 1972. Both herbicides and mechanical methods are also used in forest-thinning operations and along rights-of-way, as described earlier. Prescribed fire has a variety of purposes, among them the reduction of brush and the elimination of unwanted trees and shrubs that are sensitive to fire.

Insecticides and fungicides are generally not used in most of the forest; they find some limited use in intensively managed areas. They are commonly used in some Christmas tree plantings, nurseries, seed orchards, and urban and suburban situations.

The Congress declared almost 30 years ago that the policy of the U.S. government would be to cooperate with states, territories, and private timber owners in the retardation, control, suppression, and eradication of incipient, potential, or emergency outbreaks of destructive insects and diseases on or threatening any forest land, *irrespective of ownership* (Forest Pest Control Act, P.L. 80-110, approved June 25, 1947). Private forest-land owners, under this law, can share the cost of pest control with the federal government. This policy has been extremely important, as it places a major responsibility for decision making regarding forest pest control on the USFS. The mix of government and private pest-control responsibilities is unique to the forestry sector.

Insecticides have been used occasionally by the USFS on very large blocks of forest land, including both commercial and noncommercial, intensively and extensively managed land areas. Their use has generated considerable controversy in recent years; however, relatively few such operations have been carried out. Very few species of insects have been the object of large-scale spraying operations, and those operations have been concentrated both regionally and in terms of chemicals used.

Table 3-1 shows the approximate acreage sprayed for particular pests (including those projects supported in part or in full by the USFS) between 1945 and 1974. Only the western and eastern spruce budworms and the gypsy and Douglas fir tussock moths have been subject to insecticide sprays on over a million acres during the 29-year period. Treatments against these four species account for 96 percent of the sprayed acreage.

Table 3-2 shows the different aerially applied materials that have been used to control forest insects. DDT was the material used on 84 percent of the acreage that has been aerially sprayed since 1945. The peak year for the aerial spraying of DDT against forest insects was 1957, when nearly 5 million pounds were used. From that year, the total generally declined. No DDT was used between 1968 and 1974, when a single emergency use was approved to

TABLE 3-1 Forest Acreage Sprayed for Particular Pests: Cumulative Total for United States, 1945-1974

Insect	Acreage (1,000 acres)[a]	Percent of Total
Western budworm	12,816	41.4
Gypsy moth	12,684	41.0
Spruce budworm	3,020	9.8
Douglas fir tussock moth	1,260	4.1
Pine butterfly	255	0.8
Pitch-pine looper	239	0.8
Fall cankerworm	132	0.4
Jack-pine budworm	123	0.4
Western hemlock looper	104	0.3
Saratoga spittlebug	102	0.3
Great basin tent caterpillar	54	0.2
Pine tussock moth	41	0.1
Elm spanworm	24	
Blackheaded pine sawfly	18	
Western blackheaded budworm	16	
Redheaded pine sawfly	14	
Saddled prominent	14	
Lodgepole needle miner	13	
New Mexico fir looper	10	
Pine reproduction weevil	6	
Modoc budworm	6	
White fir needle miner	5	
Grasshoppers	4	
White fir sawfly	3	
Arkansas sawfly	2	
Eastern hemlock looper	1	
Fall webworm	1	
Forest tent caterpillar	1	
Larch sawfly	1	
Loblolly pine sawfly	1	
Lodgepole pine sawfly	1	
Southwestern pine tip moth	1	
"Utah" cankerworms	1	
Virginia pine sawfly	1	
European pine sawfly	1	
TOTAL	30,974	

[a]These figures may not represent actual acreages, as the same land may have been treated more than once in different years. See Appendix D for breakdown by year and target species.

SOURCE: Compiled by Charles Sartwell, USFS, Corvallis, Oregon, and David Alligood, USFS, Washington, D.C., from Forest Service data. October 1973.

TABLE 3-2 Materials Aerially Applied to Suppress Forest Insect Populations, Cumulative U.S. Total, 1945-1974

Material	Area (1,000 acres)[a]	Percent of Total
DDT	26,135	84.4
Carbaryl (Sevin)	2,338	7.5
Zectran	1,563	5.0
Malathion	380	1.2
Trichlorfon (Dylox)	253	0.8
Fenitrothion (Sumithion, Accothion)	222	0.7
Bacillus thuringiensis (*B.t.*)	23	
Douglas fir tussock moth virus	16	
Gardona	14	
Lead arsenate	9	
Methoxychlor	7	
Phosphamidon	7	
Dimethoate (Cygon)	2	
Propoxur (Baygon)	1	
Giant basin tent caterpillar virus	1	
TOTAL	30,974	

[a]For breakdown by year and material, see Appendix E.

combat the Douglas fir tussock moth, involving some 600,000 acres on a single-use basis.

The third insecticide on the list, Zectran, is the only registered chemical for use on spruce budworm, but production of Zectran was discontinued in 1974 because the market was too small to support the cost of its manufacture. In 1975, Zectran applied at the rate of 0.15 lb/acre on 430,000 acres in northern Maine produced excellent spruce budworm control, with a 95 percent kill of overwintering larvae. This application required only 8 days, using approximately half the aircraft normally employed in such spray operations. Zectran was also one of four candidates tested in 1972 for control of tussock moth, but because of its short residual time (normally less than 2 days) and the protracted hatching period of the tussock moth larvae, a single application achieved only a 60 to 80 percent kill. Nevertheless, these tests were inconclusive, and proper tests might show that Zectran is a legitimate substitute for DDT in the control of this insect. However, Dow Chemical, the sole producer of Zectran, has decided that it cannot produce a chemical that, at present, has such a limited market--it is not registered for tussock moth and gypsy moth control, for example, and is not widely used in agriculture, which is the major market for most commercially produced pesticides.

The USFS is quite concerned about the loss of Zectran; the agency has invested approximately $5 million in 5 years of tests of this insecticide; its short residual makes it more environmentally acceptable than many other chemicals, and it has proven extremely effective in control of a major forest pest, the spruce budworm. It would appear that unless a pesticide has a substantial agricultural market, it will not be produced commercially without some form of government subsidy, and environmentally acceptable pesticides may be lost and unsafe substitutes used because of the economics of market size.

Other Control Methods

Biological and behavioral manipulation, including the introduction and augmenting of parasites, predators, and pathogens to regulate populations of pest organisms, plus any uses of pheromones or chemical and physical attractants to influence population behavior, is a marginal operational strategy at present, despite its great promise and interest. To date, such treatments are still in the realm of research or small-scale development tests, and such techniques are directed toward a very limited number of organisms. Perhaps the major success has been the use of nuclear polyhedrosis viruses against specific sawflies. *Bacillus thuringiensis* preparations, which have features of both chemicals and organisms, find some use in controlling gypsy moth defoliation and have given encouraging results in experimental trials against the forest tent caterpillar (Abrahamson and Harper 1973). Active research is being carried out on insect pheromones; these have been used in monitoring pest populations, and they may become useful in mass trapping efforts and in male confusion procedures.

Also largely in the experimental stage is the concept of "integrated pest management," an approach which attempts to maximize natural controls of pest populations by supplementing them with other pest control procedures when necessary. Any of the above techniques or combinations of techniques may be used in order to affect the pest or potential pest adversely and aid the natural enemies of the pest. This is a promising approach and appears to have wide applicability. Its adoption, however, in forestry as elsewhere, will require that many technical and attitudinal barriers be overcome.

We examine the above techniques and the ways that they are used in some of the forest-management situations exemplified in Figure 3-1 in more detail in the following pages. The categories that we have used are defined as follows:

Noncommercial forests, as defined by the USFS (including both unproductive land and land withdrawn for wilderness or recreation uses)

Extensively managed forests, i.e., those that are managed with low to moderate intensiveness (including multiple use situations on national forest land)

Intensively managed production forests, i.e., those that are intensively managed for the production of timber, paper pulp, and so on (primarily lands owned by large private industries)

Christmas tree plantations

Seed orchards

Forest tree nurseries

Suburban and urban "forests"

These categories are not exhaustive, nor are they discrete units in terms of the degree of effort put into pest control and forest management. Noncommercial forest land, for example, goes largely untreated. Some of the land in that category, however, is used in recreation areas and thus is quite intensively managed, although the objectives are quite different from those for a production forest. Christmas tree plantations may be carefully tended with enormous technological input or casually planted and harvested with virtually no care.

Despite these limitations, the management objectives applied to any forest situation are almost always the major factors in determining the type of pest-control activity that is conducted. Therefore, we have given those objectives explicit consideration.

PEST CONTROL PRACTICES IN SPECIFIC MANAGEMENT SITUATIONS

Noncommercial Forest

About one-third of the nation's forest land is "noncommercial," defined by the USFS as unproductive forest land incapable of yielding crops of industrial wood because of adverse site conditions, plus productive forest land withdrawn from commercial timber use through statute or administrative regulation. The total area involved amounted to 253,850,000 acres in 1970 (USFS 1972). This area includes the following categories:

Productive reserved (17,236,000 acres)--productive public forest land withdrawn from timber utilization through statute or administrative regulation.

Deferred (2,713,000 acres)--national forest lands that meet productivity standards for commercial forest but are under study for possible inclusion in the wilderness system.

Unproductive (233,901,000 acres)--forest land incapable of yielding crops of industrial wood because of adverse site conditions. Sterile or poorly drained forest land, subalpine forests, and steep rocky areas where topographic conditions are likely to prevent management for timber production are included in this category. Almost half of this acreage is located in the interior of Alaska (150,798,000 acres). Part of this area will presumably soon meet the standards for commercial forest land.

The "productive reserved" and "unproductive" categories include the National Park System (30.5 million acres), at least that part that is forested, some 80 percent of the total; all of the wilderness areas (10.7 million acres in 66 units); and primitive areas (3.8 million acres in 23 units).

Clearly, "noncommercial forests" include a great variety of forests in terms of pest management objectives. Noncommercial land with no apparent usefulness to man may lie untouched, while recreation lands may range from wilderness areas only occasionally used by man to well-manicured campsite woodlands, where management may most closely resemble that of urban and suburban parks.

Pest control measures have been minimal on the rather large portion of noncommercial forest land that is virtually unmanaged. Indeed, theoretically there could be no pest control in such areas, as there would have to be some management goal in order to define any organism as a pest.

However, there are undoubtedly areas where otherwise unmanaged stands have been treated for pests, although such specific information is not generally included in project publications. In northeastern Canada, for example, tracts of unmanaged and previously unlogged land have been aerially sprayed with insecticides to kill the spruce budworm. Often the rationale for treating such areas is the need to destroy foci of infestation or to halt the spread of the pest insect into commercial areas. This is a concept that has not been adequately studied, and evidence is contradictory as to whether such movements really occur (Blais 1973). Unfortunately, since many of these areas are inaccessible except by air, there is little in the way of evaluation except for aerial interpretation of defoliation, which is almost always an inadequate means of evaluating actual population movement.

In addition, careful consideration should be given to unwanted effects of pesticides, especially with respect to potential pollution of groundwater draining into rivers and estuaries. Aerial spraying of DDT to control the spruce budworm in New Brunswick, Canada, caused extensive mortality to young salmon and their food organisms in the Miramichi River (Elson 1974, Kerswill 1967). Initially (1954-1955) the DDT sprayed on the surrounding forests killed over 90 percent of the underyearling fish and 50 percent of the salmon parr. Projecting these damage figures to the Pacific Northwest, for example, presents a potentially serious hazard that calls for extremely careful evaluation of the use of pesticides on lands that drain into important fishery streams and rivers or into estuaries, which are the nursery grounds for over 50 percent of the world's commercial fish catch.

Another such control program was conducted against bark beetles in the late 1940s and early 1950s in the Rocky Mountain region. An epidemic of spruce beetle (*Dendroctonus rufipennis*), initiated by a storm that caused extensive windthrow, killed a large number of Englemann spruce (Massey and Wygant 1954). A large amount of money was spent on treating infested trees with ethylene dibromide. The efficacy of the insecticide in the control of bark beetles is not established (Dahlsten: in press): the trees killed by bark beetles can be felled and treated to kill the brood in the phloem-cambium region (the bark), but problems of detection, timing, and coverage have greatly hindered efforts to control whole populations of the beetles (Wickman and Lyon 1962). Aside from these technical uncertainties, the decision to expend substantial funds for a control effort in a noncommercial forest may not have been justified. Outbreaks of the spruce beetles have occurred only at long intervals (approximately 50 years). In addition, the beetles may well play an important ecological role in the maintenance of Engelmann spruce stands (Miller 1970).

In many cases benefits of pest control on noncommercial sites are minimal because of the economics of such situations. The feeding of the insects, too, is not necessarily detrimental in these areas. The benefits of natural thinning or opening up of stands by defoliators or bark beetles should be more fully explored (Waters 1970).

Pests other than insects have received virtually no attention on most noncommercial forest lands. Where timber is not the object, weeds are not a problem, and the same is true for diseases.

Where noncommercial stands are used for recreation, the management goals are quite diffuse, and pest-control decisions become extremely complicated. A "pest" of any sort may be beneficial from another viewpoint. For example, a dead tree or "snag" that may be a safety and fire hazard also provides shelter and nesting sites for various small vertebrates that are valued for aesthetic reasons as well as for their role as predators of insects.

Two defoliating insects that have received a great deal of publicity have been treated relatively extensively in noncommercial stands, including all types of recreation areas. The gypsy moth (*Porthetria dispar*) in the eastern United States and the Douglas fir tussock moth [*Orgyia (Hemerocampa) pseudotsugata*] in the western United States are defoliators in both commercial and noncommercial forests. Both of these insects are discussed in Chapter 4.

Insect control in national parks has been confined largely to chemical control programs for bark beetles in high-use areas such as campgrounds. Other control efforts have involved the lodgepole needle miner, *Coleotechnites milleri*, a defoliator that has been treated in Yosemite National Park in California. Several applications of chemical insecticides have been made. Attempts in 1949 and 1953 using DDT were ineffective. Malathion was used to control adult moths in 1959 (when one application was used on the larvae as well), 1961, and 1963 (Struble 1973). The rationale was to preserve certain areas of scenic forest that had a good chance of recovery. Roadsides and campgrounds were sprayed to keep the forest green in these highly visible areas. During the three control years, a total of 9,000 to 10,000 acres were treated. The control program was judged to be successful. However, in the upper areas of the park, where the infestation has been heavy over the past 15 to 20 years (the Tuolumne Meadows and surrounding area), and where the forest was not sprayed, the number of user days never declined because of the outbreak. Indeed, the "ghost forest" (dead standing trees with a live understory) that has been created in the area through the interaction of the needle miner and the mountain pine beetle, *Dendroctonus ponderosae*, is an attraction in itself. The large numbers of young lodgepole pines beneath the dead trees makes it anything but a scene of devastation. In this case, aesthetics was defined by the control decision makers and not by the users of the park. As with the spruce beetle and Engelmann spruce, the lodgepole needle miner seems to play an important role in maintaining stands of lodgepole pine, a role similar to that of fire in stand establishment (Telford 1961, Fowells 1965). Education of the public about the ecological processes taking place might be a better approach to the problem of pest outbreaks in recreation areas, especially as there is such a diversity of objectives in the use of these areas.

Pest control efforts of any type in low-use recreation areas are not easily justified, and in general these low-use areas, including wilderness and primitive areas, have not been subject to control efforts. At times the potential for spillover of pest problems into commercial areas has been used to justify control efforts on recreation lands as well as on other noncommercial land. Care should be used in predicting such effects, as the dangers of their occurrence may be minimal.

Weed and disease pests in recreation areas generally are not controlled because the damage they may do in such areas is usually of little or no consequence. Vertebrates of all sorts are often highly valued on recreation lands, although the lack of predator pressure may allow certain animal populations to increase beyond the carrying capacity of the land. In such cases, increasing the human predator pressure, by way of longer hunting seasons and similar measures, is an accepted form of wildlife management.

In conclusion, justification of pest-control projects in noncommercial forest land is often impossible because of the lack of real benefits associated with such efforts. Where recreation is an important activity, the goals of control are difficult to attain because of the many diffuse purposes and values of the users of these areas. When control is to be considered, the decision should be made in light of the diverse interests concerned with the resource. Full advantage should be taken of the educational opportunity in noncommercial forests with respect to biology, ecological processes, and a demonstration of an holistic approach to pest-control decision making.

Extensively Managed Forests

Extensively managed forests are those commercial forest lands from which benefits are derived with a minimal or relatively low input by the owners and managers of the land. The term also implies a planning unit size that is larger than optimum in terms of effective management. This category cannot be described quantitatively, as there is a continuous gradation in the intensiveness of management activities, including pest control, on forest lands. We can, however, describe the sort of land that fits our definition of the extensively managed forest and discuss current pest control in those areas.

The category includes forests on relatively poor sites that grow slowly and meet only the minimum standards for commercial forest status. It also includes many small privately owned woodlots on excellent sites. In either case, the forest may receive very little attention from the owners. In one, the low value of the products does not permit economic investment; in the other, a perfectly intelligent market decision or preference of the owner may work against intensive management practices.

The objectives and practices of nonindustrial private woodlot owners have been of concern to those who wish to increase timber product output from such lands for many years (see PAPTE Report 1973, Chapter 8 and Appendix I, for a detailed analysis of the problem and various methods that have been used to attempt to deal with it). It is generally agreed that the average level of management practiced on such lands remains far below that of industrial lands, despite the existence of various state and federal incentive programs. A primary aim of incentive programs is to promote stand improvement through selective thinning and brush control and to encourage effective regeneration of cutover lands. Thus, an improvement in the effectiveness of such programs in terms of pest control would probably have the most effect on the control of weeds. Insects and diseases may be controlled by any forest owner on a cost-share basis under the Forest Pest Control Act of 1947 (see Chapter 4). Many small woodlot owners, however, do not take advantage of such programs. In general, their land is little managed and pests are not controlled. It should be understood, however, that some lands in this group are under intensive care, as is indicated by the number of small owners participating in the National Tree Farm Program.

Extensively managed forest land also includes a portion of the national forest lands, where "multiple use" is an explicit objective. About half of the total national forest area is classified as commercial forest land. Compared with other ownerships, this land shows the largest difference between current timber growth and growth potential of the land, primarily because of the large stands of old-growth timber on western national forest lands, where growth has slowed in mature trees. In such areas, national forests are not being harvested and replanted nearly as quickly as forest industry lands, in part because of a lack of funds, in part because of the way that allowable cut standards are set to ensure sustained yield, and in part, too, because of the principle of multiple use, under which management for timber production is not always given first priority.

On both national forest lands and extensively managed nonindustrial private lands, pest control is likely to be confined to occasional measures against insects such as the spruce budworm and bark beetles. Such insects may kill trees over very large acreages in outbreaks, and control measures may be used to protect the entire forest area.

Although a large percentage of the present commercial forest acreage is extensively managed, this situation is changing, especially on the larger industrial holdings, but also on public lands and, to a lesser extent, on nonindustrial private lands. As the demand for timber products increases and the supply of old-growth sawtimber dwindles (in the Pacific Northwest), forest owners will use their resource more intensively, and--one hopes--with increasing care. In the pest-control area, this should mean increasingly selective application of control measures as well as increased capital and technological inputs.

Intensively Managed Production Forest

As previously pointed out, intensive forest management has two components. It most commonly means that high levels of technological effort are expended to insure prompt establishment, rapid growth, and protection of a wood crop. In principle, similar levels of effort may be directed toward wildlife production, recreational values, or restoration of a damaged watershed. Here, however, we are concerned only with forests managed for timber production.

The second component is the minimum size of the distinctive land unit that is recognized for planning purposes. The smaller the unit designated as different from its surroundings, the more likely it is to receive appropriate treatment, even though the total effort applied to the entire forest may not change. Logically, recognition of planning units as needing intensive management should precede intensive technological input, although such has not always been the case. The operational unit for treatment, however, need not coincide with the unit for planning--just as a farmer's field remains the operational unit even though the planned treatment takes account of the diverse soil types within it.

The expenditures required for intensive management are warranted only where soil and climate allow high productivity. Owners and managers are willing to make such expenditures only when land costs are high, competition for raw wood supplies is great, and harvest and manufacture operations can be carried out efficiently. Another limitation on the extent of intensive culture is availability of investment capital. Little of the early cost is recoverable in less than two to three decades, and meanwhile the young forest is exposed to a variety of natural hazards. At present, intensive management in the sense used here is applied to significant areas only in the South and Pacific Northwest.

The exact acreage under highly intensive management is unknown. However, in 1973, the forest industries in Washington, Oregon, and nine states of the lower South planted 790,000 acres, some 80 percent of the total planted in those states (USFS 1973). It is likely that more than three-fourths of this area would be classed as intensively managed. The acreage of industry land sown or planted totals approximately 10 million acres, but by no means all of this qualifies as successful intensive culture in the present sense of the term. However, appropriate additions from other land--federal, state, and other private--would mean at least 5 million acres of plantation under intensive culture, together with an unspecified large area of naturally sown stands.

Intensive management should not be equated with plantations and plantation management. In the past, much tree planting created or restored forests that were then managed extensively or not at all. On the other hand, increased technological inputs can be made to native as well as planted forests, at any stage before maturity. Native forests on highly productive sites often respond profitably to more intensive cultural treatments. The essence of intensive management in respect to pest problems is, first, that the operational units are small enough and accessible enough for close surveillance; and second, that the high monetary values involved in establishing and maintaining the stand impel the manager to respond promptly to threats of loss whenever and however these are perceived.

The actual operational approaches to pest control under intensive management can be considered in terms of the manager's strategies discussed earlier. All of the approaches described there, from the acceptance of injury through the use of pesticides, find some application in intensive forest culture.

The levels of damage that are economically acceptable are determined by the investment costs in these forests. Damage in the form of brief delays in initial establishment or fractional losses of yearly wood production in occasional years are likely to remain tolerable, provided that stocking, wood quality, and vigor are not reduced appreciably. Feasible means of preventing such loss or delay may either be unknown or involve intangible risks or diversion of skilled effort that negate otherwise favorable cost-benefit ratios. Many persistent but minor sources of loss and some conspicuous but infrequent episodes fall within this category. (An example is the needle cast epidemic, attributed to two common native fungi, *Lophodermella cerina* and *Hypoderma lethale*, that in 1971 affected pines over 54 million acres in the southern United States. Such an epidemic was previously unknown in this region and has not recurred.)

On the other hand, those pests that prevent full stocking or cause persistent defects in young stems or continuously retard growth will result in intolerably large cumulative losses and thus demand control. In such situations, control is often by chemical means and in anticipation of damage rather than after its occurrence. Thus, there will sometimes be acute conflicts between strategies of "control" and "tolerance" and their respective proponents.

Certain pest problems can be avoided by either not planting a susceptible species in habitats where damage is likely or by modifying habitat features that predispose to injury. In principle, intensive culture provides greater opportunities to so adjust and manipulate. For example, surveys indicating zones of high and low hazard from fusiform rust can guide decisions on whether to plant the highly vulnerable slash pine, or the susceptible loblolly pine, which is somewhat less damaged by infection, or even the slow-starting but more resistant longleaf. In addition, it is known that treatments that hasten early growth commonly increase the probability of infection, so some trade-off between initial growth rate and infection hazard may be advantageous.

In actuality, the opportunities to avoid pest damage are limited by the relatively small number of species now known to be adaptable to intensive culture, by trade-offs with other risks, and by incomplete knowledge about many pest-habitat interactions. In the example cited above, the effective choice is often between slash and loblolly, since longleaf is currently not used in short-rotation culture, except where wildfire risk is high.

In intensive culture, avoidance will increasingly take the form of genetic selection for resistance with otherwise promising economic species, rather than substitution of species. There is a good reason to expect that over the next three decades genetic selection will be a major factor in reducing or rendering manageable damage from some important pests, especially fungi, although it undoubtedly will be ineffective against others. Selection for growth and site adaptability may also render additional species attractive to intensive culture and so increase crop diversity. Genetically improved stock already constitutes more than half of the current planting on some industrial holdings.

The first major task of tree-improvement research has been to establish the variability of species populations over their geographic and ecological ranges and the suitability of specific relocations of plants from seed collected in a particular locality. For several species, large differences in climatic adaptability, growth characteristics, and susceptibility to pests are now well known through extensive experimental plantings. The first such trial with loblolly pine, for example, showed a range of 6 to 32 percent fusiform rust stem cankers and 14 to 6 cords of wood per acre at age 15, depending on the geographic source of the seed (Wakeley 1954). As a result of many such studies, seed-collection areas appropriate to planting in specific locations now have been mapped out.

This research continues, but two other approaches have been directed more specifically to the opportunities and needs inherent in intensive culture. Outstanding phenotypes have been selected on the basis of several criteria, including apparent resistance to pests. These have been established in seed orchards and simultaneously subjected to progeny testing. Maintenance of genetic heterogeneity has been emphasized by mixing numerous clones within each orchard. Seed from these first-generation orchards already makes up a large fraction of the total loblolly pine and Douglas fir used in planting for intensive cultures and will soon dominate to the exclusion of wild seed. Second-generation seed orchards will make use of further selection that emphasizes resistance to major pests and responsiveness to the modified environments created by intensive culture. In the southern United States, mass screening and other procedures are now concentrated in a search for resistance against fusiform rust, a major disease of young loblolly and slash pines under intensive culture conditions there.

Loss of resistance due to mutation or population changes by the pest is certainly a concern. One evident safeguard is maintenance of a high degree of genotypic variability within all improved selections. This is made easier because the practical requirements for phenotypic uniformity, apart from height growth and wood density, are less stringent than with most agricultural and horticultural crops. Another avenue is through multigene rather than single-gene sources of resistance, especially to rust fungi. A further consideration of importance here is the time span of vulnerability to a given pest. The major economic losses caused by fusiform rust, for example, result from stem or near-stem cankers formed in the first 5 to 8 years after planting in intensive cultures (assuming protection in the nursery). Loss of resistance after such a period would have very modest consequences for stands of this age or older. A similar span of resistance would suffice for several other pests of juvenile stands.

Lands under intensive management are well adapted to sanitation procedures for pest control because of accessibility and close control over stand conditions. No matter how sound and acceptable the biological basis, however, sanitation per se may be wholly ineffective in the face of major outbreaks unless coupled with other measures, with survey and reporting systems, and with a capacity for prompt action. The need for such combination is exemplified in the southern pine beetle story (see Chapter 4, p. 120).

Direct control of pests by chemical or mechanical methods is presently restricted largely to two situations, even under intensive management: before or during the early establishment period and during intermittent outbreaks that threaten excess damage or mortality. Few wood crops are sufficiently valuable to warrant application of protective chemicals on an annual or more frequent basis throughout the life of the forest.

Intensive management generally means some form of even-age management system, both because of the growth and regeneration patterns of some of the major commercial species, such as the southern pines and Douglas fir, and because of the economics of production and harvest. At some stage, then, the existing forest is harvested, in small or large units, and a new forest is established.

The particular measures used to establish the new forest differ according to species, soil, and climate, but the general features are those described in our earlier section on site preparation. Natural reproduction is still used and indeed cannot be avoided in some locations but, increasingly, nursery-reared seedlings (often of genetically improved stock) are planted at spacings that allow unimpaired early growth and facilitate later cultural and harvest operations. The pest-management emphasis at this stage is on protecting the vulnerable seedlings from competing vegetation, insects, and vertebrates. Costs of site preparation and planting commonly range between $75 and $200 per acre.

Young forests regenerated naturally under favorable conditions are often excessively dense and must be thinned early to avoid several decades of slow diameter growth before the trees reach economic sizes. In the Pacific Northwest, stands may be thinned from 2500-3000 down to 450 stems per acre. Despite attempts to mechanize, such precommercial thinnings involve a high labor input, with costs up to $50 per acre. Hardwood plantations nearly always require control of woody and herbaceous vegetation in the first 2 to 3 years after planting. This may be achieved either by cultivation or herbicide application. Thus far, the total area involved annually amounts to no more than several thousand acres. With conifers, postplanting control of competing vegetation is commonly not required when site preparation has been intensive. Currently, however, the value of postplanting weed control to hasten early growth is under investigation, and it is possible that cultivation and herbicide use will become more widespread.

In general, fungi present no major handicaps to successful establishment of the species now used in intensive culture, with two exceptions: the already-mentioned rust fungi of southern pines and the root rots such as *Fomes annosus* carried over from the previous forest. As additional species are brought under intensive culture, it is likely that additional fungi, especially those causing leaf and shoot diseases, will become important.

Given adequate control of competing vegetation, the greatest remaining hazard to uniform establishment of seedlings is a certain number of insects and mammals.

None of those animals are specialized feeders, but the seedlings are small and widely spaced; even incidental feeding that would be inconsequential on large plants becomes highly destructive. The widespread Pales weevil and the locally important Texas leaf-cutting ant, for example, can essentially destroy a prospective forest within a few weeks or even days after planting. The same is true of damage by vertebrate pests, except that the trees remain susceptible for a longer time, often until they have grown beyond the effective reach of the animals. The hazard varies from none to very high, depending on the region, habitat, and tree species involved. Direct seeding is not commonly used in intensive culture, but mice and birds are almost invariable problems when it is. Where hazard is high, the direct control measures discussed in the previous section are necessary and are often applied before damage is apparent.

As a plantation grows into the sapling stage, shoot injuries by either insects or fungi have serious consequences, because the potentially most valuable portion of the trunk is being formed. Defoliation is sometimes of importance then, but is likely to become more so later, in the "pole" or "small sawtimber" size classes, when the importance of young shoot damage declines. At these stages the hazard from root rots such as *Fomes annosus* or *Poria weirii* also becomes of increasing concern.

Factors of cost, equipment, and scale of operation provide a continuing check on any casual use of direct treatment, but, nevertheless, the area treated annually seems certain to increase rapidly over the next two decades as the area of intensively managed plantations expands and includes additional species, and as damage from previously unimportant insects and diseases is encountered.

Active techniques of biological manipulation to decrease pest problems may well increase in importance in the future. These may include the use of mixtures of species, as well as the application of specific microbial agents to control forest insects and diseases. The former is still largely in the realm of theory and experiment. Of the latter, as mentioned earlier, *Bacillus thuringiensis* has been used with some success against the gypsy moth and forest tent caterpillar. Also, foliar sprays of polyhedrosis viruses used experimentally have given effective control of some sawflies, indicating possible commercial use in the future. No such techniques are yet of operational importance in intensively managed forests.

There are several variations of the pattern indicated above. A number of less conventional methods of culture are being explored. Some would produce small-diameter wood for bulk harvesting and chipping; others would produce standard pulpwood through short, no-thin, single-harvest rotations; still others, a mixture of products through longer rotations with intermediate thinnings before final harvest. All, however, tend to use relatively short growth periods (i.e., 5 to 50 years), selected species or selected genotypes within species, and systems of efficient production that extend from seedling establishment to harvest and product utilization. The consequent changes in pest problems and outlooks toward control are great.

The techniques of intensive management create certain problems and reduce others. The insects and pathogens associated with particular species, with planting, with juvenile stages, and with rapid growth rates take on greater importance in an intensively managed than in a mixed natural forest. With short rotations, loss of only a small percentage of widely spaced seedlings affects ultimate yield. Preplanting or postplanting tillage, thinning, and mechanical operations of all types present new opportunities for certain types of biotic and abiotic injury.

The extensive areas of single species of uniform age created with present intensive practices can be expected to be more vulnerable to certain pests. In the United States, nearly all tree species now important in intensive culture are native to the region where they are grown.* Most of the species included in our generalization occur naturally, at times, in even-age stands or very simple mixtures. Nevertheless, it is apparent that plantation management, with its repeated planting and harvesting of uniform crops represents a different ecological system from that in which pest-host relationships evolved. The possibilities for pest problems are thus likely to be greater in plantations than elsewhere.

The increasing use of tree species genetically selected for various characteristics carries with it certain potential dangers. Genotypes may be selected for desirable

---

*Two exceptions are slash pine and red pine, which have been planted well outside their natural ranges. Also excluded from the generalization are many forest plantations in the Northeast, where European species and native species outside their range have been planted, but only a few of these are intensively managed according to our definition.

characteristics (such as fast growth) but, despite testing, may later prove susceptible to particular insects or diseases. It appears, however, that the chance of narrowing the genetic bases of most tree species to the point where the selections risk general and extensive susceptibility to particular pests is quite small. For the immediate future, the forest geneticist is concerned with the early stages of selection from heterogenous native populations that are usually of widespread geographic and ecologic occurrence and are unaltered by previous domestication.

One area where problems of genetic uniformity occasionally arise is in the use of species that are propagated vegetatively. With cottonwood, the danger of excessive uniformity was specifically countered by the release of 14 select clones for commercial production, with the recommendation that they be used in mixture (Johnson 1972).

Systems of intensive management also decrease the importance of certain other types of pest problems. For example, a number of pest problems responsible for slow net growth and mortality in most forests may become less important under intensive culture. The enormous losses of seeds and newly germinated seedlings that are responsible for the unpredictability and hazards of natural regeneration systems are avoided through the use of seed orchards and nurseries. In a sense, the pest control problems of seed and new seedlings are consolidated into those very small areas of specialized culture, although the newly planted seedlings still must be carefully tended in the forest. Brush and woody weeds, probably the largest single cause of inadequate stocking in natural forests, are controlled, though seldom eradicated, by intensive site preparation. The wood rots responsible for great cumulative losses, cull, and product degradation in older forests are almost entirely avoided by short rotations. Similarly, young vigorous stands are generally less susceptible to catastrophic bark beetle epidemics, which typically originate in overmature trees or those stressed by drought and other agents (although some species of root rots and bark-feeding insects readily damage young trees as well as old).

If smaller planning units as well as technical and capital inputs are used as criteria for judging true "intensive management," then it should be possible to actually reduce the need for large-scale chemical control treatments in such forests. The ease of access to the stands should facilitate thorough evaluation work and prescription of spot treatments where they are necessary. The very high productivity of forests under intensive management should provide incentives for including pest management within their culture.

Christmas Tree Culture

Although the area devoted to Christmas tree culture is relatively small, it deserves brief discussion for two somewhat conflicting reasons. First, the very high cultural inputs, dependence on frequent application of pesticides, and wide use of species not indigenous to the site more closely resemble horticultural practice than the intensive culture of wood crops described above. Second, the methods used exemplify numerous pioneering attempts to grow trees under nonnatural conditions; the pest problems encountered and the methods developed to promote rapid growth in the face of continuing pest impacts will sometimes apply to other systems of forest tree culture.

The total area in Christmas tree culture is estimated roughly at 450,000 acres (A. T. Davidson, personal communication). However, the intensiveness of culture varies from fully commercialized systems with a high level of material input and high value of output through all degrees of tended plantations and selected wild stands. Many plantations classed as Christmas trees in the past have in fact simply yielded one partial crop, with only moderate inputs, while serving as a transitional stage between the abandoned fields in which they were planted and young forests of uncertain quality and value. In contrast, many present-day commercial operations are systematically harvested and replanted as a crop, with large annual investments in shaping, fertilization, vegetation control, and pest management.

Although commercial producers dominate the major markets, the extreme seasonality and other peculiarities of demand and distribution allow many small and less commercialized growers to stay in businesses. Thus the area of these plantations in highly intensive culture is probably much less than half the estimated total. In 1973, plantings exclusively for Christmas trees totalled 33,700 acres. If all remained fully successful, which is most unlikely, and were managed on a 10-year rotation, which is overly long as a national average, the total area would be only three-fourths of the previous estimate, and the area under fully commercial management still less.

Christmas tree growers must contend with four general classes of pest problems: woody and herbaceous vegetation; browsing animals; shoot and leaf insects and pathogens; and root and stem insects and pathogens.

Commercial growers usually control woody and herbaceous vegetation prior to or soon after planting to facilitate establishment and rapid early growth. Either cultivation or herbicides or combinations may be used for initial suppression; band treatments with herbicides are useful where full soil exposure is undesirable. Subsequent suppression is by herbicides or mowing, although maintenance of continuous bare soil conditions is feasible only on highly stable, porous soils such as those in the Pacific Northwest. Selective herbicides such as 2,4-D and 2,4,5-T are widely used even in less intensive culture to reduce competition and abrasion from woody or tall herbaceous vegetation.

The problems of browsing and clipping by rodents and deer vary greatly with region, locality, and tree species. Smaller rodents, such as voles and gophers, are often controlled by poison bait or other applications not highly destructive of nontarget species. Game species must be controlled by other means. Exclusion by fencing involves high capital costs. Repellants are often recommended, but heavy and continued browsing pressures require repeated applications at substantial cost. Manipulation of food and cover on site and in the vicinity may occasionally ease control problems. For example, gopher damage can sometimes be prevented by the eradication of the animals' preferred herbaceous food plants.

A Christmas tree approaching merchantable size represents a substantial cumulative outlay in past care; its value, however, depends on a lack of evident damage to buds, shoots, or foliage when it is ready to be sold. Many insects and needle fungi of minor significance in wood-production forests, therefore, take on much greater economic importance in Christmas tree production.

Damage from a few such pests can occasionally be controlled by nonchemical means when infestation rates are low. Also, tolerance of minor infestations is increasingly advocated as less costly than full-scale control, as well as being free of secondary consequences. However, major dependence is placed on insecticides and fungicides, usually applied as sprays by ground equipment. The materials and rates are standard and comparable to those employed in horticultural uses or in forest nurseries, but timing is often critical for effective control. The latter fact, together with the considerable values involved and the general needs for diagnosis and exact technical guidance, places a premium on good advisory services. These are provided by grower associations and pesticide manufacturers and, in many areas, by the state extension services. The latter may be influential in advocating pest "management" rather than "control," including avoiding susceptible stock, tolerance of some damage, sanitation, spot rather than broadcast treatments where appropriate, and choices among effective materials and methods to minimize known side effects. Thus, for example, alternatives to chemicals toxic to bees may be recommended, or short-lived chemicals or dormant oils to avoid reducing spider-mite predators. Biological control techniques have not been much used in Christmas tree plantations.

Several insects damage the roots or stems of recently planted trees; these are readily controlled chemically, sometimes by root dips with persistent insecticides. Chemical controls are likewise effective against a number of other insects. Occasionally, stem rusts and branch fungi such as *Cytospora* cause serious damage. Either preventive sprays or sanitation in specific cases provides substantial control. *Cytospora*, for example, is easily spread by pruning tools in wet weather. Potentially, the most difficult problem is root disease caused by soil-inhabiting fungi. Root damage caused by *Phytophthora cinnamomi* has caused up to 5 percent loss in Frazer fir plantations in North Carolina, for example. Such fungi are often exceedingly difficult to control except by soil fumigants. As yet, however, they have not proved to be major concerns.

Seed Orchards

Management of forest plantations for seed production involves only a very small percentage of the total forest area. In 1974, seed orchard acreage (private, industrial, state, and federal) totaled 9,263 acres of all species in all states. (USFS 1974).

The seed produced is of high value, so management practices are frequently intensive. These include measures to protect the plantations from all types of pests. Weed problems, prevalent at the time of establishment, are replaced by insects as the major pest problem when the trees reach producing age. Large vertebrates are generally fenced out, and rodents do not normally cause trouble where weeds are controlled. Diseases are numerous, but of variable importance.

Weed control programs involve conventional tillage in preparation for planting, application of triazine herbicides during the first 1 to 3 years of growth, and mowing thereafter. There is no apparent need for additional weed management tools.

Seed orchards are usually small and are often located in fringe areas between inhabited and forested areas;

accessibility and suitability for forest growth are both important criteria in site selection. Because of their proximity to large populations of normal forest insects, it is generally not possible to eliminate insect pests in a plantation without immediate reinfestation. When seed orchards are near urban centers, additional problems may be created by air pollution, which can increase the vulnerability of the trees to insects and diseases.

Insect problems are specific to individual production areas. Of most importance are the seed and cone insects. Most of these insects cannot be killed by surface applications of insecticides except during a limited portion of their life cycles. For this reason, timing is critical in the use of contact sprays. Systemic insecticides are used to some extent, but the results obtained are variable.

Insect problems can sometimes be avoided through cultural manipulations. Insecticides are also used in seed orchards but not as intensively as in fruit orchards. Spot treatments of individual infestations are common, and broadcast sprays are also used occasionally.

Numerous diseases occur in seed orchards. The most serious may be the fusiform rust of southern pines. Some diseases are important because of the specific management practices used in seed orchards, but these have not been well identified. Copper sprays and other fungicides are occasionally used in seed orchards.

Disease resistance may be of increasing importance in seed orchards in the future. Breeding programs for resistance to disease have been in progress for white pine blister rust for at least 20 years; programs for other diseases have been in progress for lesser periods of time. Progenies have been identified that are resistant to various diseases, and it is likely that seed production areas will increasingly furnish significant amounts of disease-resistant stock. Some questions regarding gene pool management will need to be resolved before breeding programs actually constitute a major part of disease management.

Pest control efforts in seed orchards are necessary for continued "tree improvement" and are economically justified because of the high value of the product. The diversity of pests and their localized nature make a variety of tools necessary, including combinations of insecticides and cultural methods.

Forest Tree Nurseries

The acreage of forest tree nurseries in the United States is similar to that of seed orchards--about 9,000 acres.

Despite the small area, the value of seedlings produced is very high, and pest problems are therefore costly. Major nursery pests include weeds, fungi, insects, and nematodes, in decreasing order of importance. Pest control is practiced as intensively in nurseries as in most other crop-producing areas, and control practices are changing as new technologies and higher wages force a reduction in the use of hand labor.

Weeds are presently controlled by both hand and chemical methods. Labor requirements for hand weeding are responsible for a large proportion of the total costs of production, although some herbicides are used. Apart from the widely used light oils, only a few chemicals are presently registered for commercial use in tree nurseries, and the selection is inadequate to control the diversity of weeds found in the highly fertile nursery beds. This problem is likely to remain significant because of the heavy liability to registrants in the event of herbicide damage and the limited market potential for registered products. Herbicides currently used are DCPA, diaphenamid, and simazine.

Fungi cause heavy losses in nursery operations. Damping-off of newly emerged seedlings is the first major problem. At later stages in the growth of the seedlings, various root rots are often important. Most of the common root diseases and nematodes can be kept under control with soil fumigation prior to sowing. The chemicals most commonly used are methyl bromide and chloropicrin or similar materials. Foliage diseases, such as the red band needle blight of pines, are generally minor production problems, but some foliage and stem diseases must be controlled in the nurseries to prevent later damage in outplanted trees. Sprays of copper compounds and benomyl are occasionally used and have proven effective in arresting a number of diseases after they have emerged.

Overall, insects cause relatively minor losses in forest nurseries, although some could be very damaging in the absence of control. Some soil insects such as wireworms, and white grubs and *Brachyrhinus* weevils cause sporadic mortality, but seldom cause widespread losses. Considerable money and effort is spent in preventing the escape of exotic insects from the nursery to the forest. The European pine shoot moth, for example, requires constant vigilance in nurseries of the Pacific Northwest. Many soil insects found in nurseries are kept in check by soil fumigation and are not serious problems where fumigation measures to control root rots have already been taken. Fumigation is not always necessary, as there are cultural methods available to control some of the soil insects as well.

Seedlings in the nursery are sometimes treated with repellent sprays to reduce browsing damage soon after outplanting.

Although the acreage devoted to nursery management is small, pest control practices can be intensive. Toxic materials are used (fumigants and certain of the repellents), and there is considerable handling of seedlings by hand after treatment with various chemicals. At least one chemical used in nurseries can have cumulative effects in soil (ZIP, an animal taste repellent).

The tree nurseries, like the seed orchards, require a substantial investment to produce commodities of very high unit value, so that protection from pests becomes a necessity. Both are important beyond what their small acreage would suggest, as much of the nation's timber-producing acreage depends on their products.

Suburban and Urban Forests

The suburban forest is difficult to define and quantify with any precision. It is composed of a mosaic of wooded acres of diverse size interspersed with developed residential land. As expressed by Waggoner and Ovington (1962), it is forested land "caught within the metropolis and its web of communication." In its broadest sense, it includes almost 2 million acres of woodland in such a highly populated state as Connecticut. The urban forest is interspersed with even more intensively developed areas than the suburban; this interstitial forest includes trees growing between man's constructions as well as private wooded estates and tree-covered city parks (Detwyler 1972). The total area designated as "urban" (including "urbanized areas and other incorporated and unincorporated places of 1,000 population or more") was estimated at 34.6 million acres nationwide in 1969 (USDA 1973). The urban and suburban forests not only cover impressively large areas but are also highly dynamic. For example, in Philadelphia, the diameter of the circle of urbanization including the suburban-urban forest increased from less than 8 miles to 16 miles during 30 years (Toulan 1970). The total "urban" area increased 7.4 percent between 1959 and 1969, while the acreage of "forest land" fell slightly.

Aspects of the urban-suburban forest have been described by Clement and Nisbet (1972), Odum (1971), Smith (1969), The University of Massachusetts (1970), Waggoner and

Ovington (1962), Gill and Bonnett (1973), and Detwyler (1972). Because of its setting, this forest has unique values, involved primarily with aesthetics, wildlife habitat, and amenity functions.

Urban areas are characterized by a surprisingly large tree population. These trees are planted along streets, in parks and playgrounds, and around residential, commercial and industrial establishments. The city of Philadelphia has approximately 300,000 street trees alone and is adding 5,000 to 6,000 per year to city streets (McConnell 1973). In New York City, street trees number less than 30,000 in Manhattan, but there are approximately 500,000 in the boroughs of Queens and Brooklyn combined.

Individual trees often have considerable value (Smith 1971) in the urban-suburban situation. Methods have been developed to provide appraisal values for these individual trees. The evaluation method of the International Shade Tree Conference, Incorporated (Lewis 1970), applied to high-quality ornamental specimens growing in a prime New England location, indicates individual values as high as $3,140 for a 20-inch (diameter) white oak (*Quercus alba*), $7,070 for a 30-inch sugar maple (*Acer saccharum*), and $12,570 for a 40-inch American elm (*Ulmus americana*).

Another facet of the dollar value of urban trees is revealed by the amounts spent on removal and maintenance. In the city of New York, for example, $375,000 was spent in 1967 to remove dead trees; this does not include the additional cost of removing 770 dead elm trees killed by Dutch elm disease. In Chicago, street trees and shrubs alone total over 800,000. The Chicago Bureau of Forestry recently estimated that over 295,000 20-inch American elms will die in that city and will have to be removed by 1978 at an approximate cost of $24 million (Andresen 1973). In 1968, approximately $500,000 was spent to prune 30,000 trees in Queens and Brooklyn, New York. The city of Portland, Oregon, annually spends $15,000 to spray 600 elms to protect them from Dutch elm disease. A recent survey of municipal arborists conducted in New England and the Middle Atlantic and Central states suggested that approximately 100,000 trees were planted annually along highways and 200,000 in municipalities in 13 states. Costs for the more than 800,000 trees planted during 1968-1972 were estimated at over $14 million for purchase and over $32 million for planting (Long et al. 1973). Clearly, money in these amounts would not be spent unless the trees were considered to have appreciable value.

In view of this value, control costs associated with pest problems clearly can be justified. All indications suggest that these costs, and the amounts of pesticides used in urban/suburban areas, are extremely high. Unfortunately, we were unable to obtain any estimate, either of dollars spent or of chemicals applied. No federal agency is charged with collecting such data, and no trade or other association has attempted to gather them.

The responsibility for control of insects and diseases of urban and suburban trees is diffused. In suburban areas, a large number of professionals, semiprofessionals, and nonprofessionals (arborists, tree surgeons, landscape contractors, and handymen) are available to the tree owner on a fee basis for pest control. In urban areas, several layers of government--for example, county and municipal--may be involved. Private citizens, either as individuals or as members of groups, may assume responsibility for tree protection. Private commercial and industrial organizations are also involved. The conduct of control operations is largely carried out by commercial arborist organizations and by full- or part-time government employees. In large cities, full-time public personnel invariably assume responsibility for tree planting, removal, and protection.

Pest problems may be very significant in urban/suburban areas. Species may be planted that are not indigenous to the area, and the planting site may be poor. Both biotic and abiotic pathological stresses are encountered. Pesticides are commonly employed against pathogens. Examples are control of anthracnose leaf pathogens of eastern hardwoods with Bordeaux mixture and ferbam (Berry and Lautz 1972); of arborvitae, cedar blight and birchbrown leafspot with Bordeaux mixture; of dogwood flower and leaf blight and oak leaf blister with zineb; and many others. Cultural, nonchemical controls are also commonly employed against pathogens. Examples are control of spruce canker by pruning cankered branches, and of white pine blister rust by pruning symptomatic branches; it is necessary to completely remove (eradicate) maples with bleeding canker and mimosas with *Fusarium* wilt (Wester 1968).

Abiotic stresses receiving considerable attention include nutrient and desiccation stress. Fertilization of established trees has been a standard recommendation as a preventive or corrective measure for several tree diseases, although the practice is supported by relatively little scientific evidence (Neely et al. 1970). Numerous treatments of dry and liquid root feeding and foliar and injection feeding are practiced in suburban residential, commercial, and industrial locations. Desiccation stress (winter killing) occurs throughout temperate regions where fall and winter winds and temperatures dehydrate plants. Numerous controls to mitigate this problem include watering, mulching, covering, and spraying with wax emulsion or plastic-based compounds.

Because aesthetic values are of considerable importance in urban/suburban areas, defoliating insects may stimulate chemical control measures primarily because the pests remove the leaves from the trees, whether they actually harm the trees significantly or not. The elm leaf beetle, for example, was in the past heavily treated in some areas simply because it defoliated trees, although permanent damage often was minimal. In other cases, the presence of an insect alone is enough to initiate control efforts. For example, the spring and fall cankerworms, which actually do little damage, are often controlled as nuisance pests. The gypsy moth is among the most serious pests of suburban areas in the Northeastern United States and, indeed, is among the few forest pests that have been heavily sprayed with insecticides. Both the presence of this insect and its evident damage have stimulated control efforts near residential areas (this important pest is discussed further in Chapter 4).

Trees in the city are growing in an even more artificial environment than that of the suburbs, an environment that bears little relationship to the ecosystems in which the trees evolved. They are confronted with numerous abiotic stresses, including poor air and soil quality, all of which may increase their susceptibility to pest damage. Biotic pathogens and insect pests, once established in urban areas, may greatly expand their populations, frequently in the absence of natural controls, and cause damage of epidemic significance.

Controls applied in the urban situation are frequently intense and quite varied. Cultural and chemical controls similar to those used in suburban forest areas are employed.

An example of urban tree disease control relates to the protection of American elms from infection by *Ceratocystis ulmi*, the fungus that causes Dutch elm disease. For 300 years, American elm has been the most important shade tree in the United States. Approximately 40 percent of the 25 million elms that have been planted over the years have now been killed by the disease. At present, elms are being lost at a rate of 400,000 annually, at a cost of approximately $100 million. Table 3-3 shows recent losses of elms to the disease in some midwestern cities.

The standard means of protecting elms has been by way of an insecticide rather than a fungicide. For years, DDT

TABLE 3-3 Losses of American Elms to Dutch Elm Disease in Some Midwestern Cities[a]

City	Population (1960)	Peak No. of American Elms	Loss of Elms per Year (1970)	Program of Control by Spraying in 1970	Year Spraying Stopped
Ypsilanti, Mich.	21,000	10,000	1.0% (2,000 left)	Total[b]	--
Pittsburgh, Pa.	604,000	25,000	1.2	None	1968
Birmingham, Mich.	26,000	8,000	1.5	Total[b]	--
Evanston, Ill.	79,000	19,000	c.2.0	Total[b]	--
Flint, Mich.	197,000	62,000	2.0	Total[c]	--
Midland, Mich.	28,000	15,000	c.3.0	Partial[b]	--
Elmhurst, Ill.	40,000	15,000	3.0	Total[b]	--
Detroit, Mich.	1,670,000	310,000	3.2	Partial[b]	--
Dayton, Ohio	262,000	20,000	4.5	None	--
Saginaw, Mich.	98,000	20,000	4.5	Total[b]	--
Pontiac, Mich.	82,000	2,600	5.8	None	1969
Rochester, N.Y.	319,000	50,000	6.0	None	1960
Ann Arbor, Mich.	67,000	10,000	7.0	Partial[b]	--
Toledo, Ohio	318,000	72,000	10.0	None	1964
Coldwater, Mich.	9,000	11,000	10 to 20	None	1963
Milwaukee, Wisc.	741,000	127,000	? (62,000 left)	None	1966
Battle Creek, Mich.	44,000	4,700	? (400 left)	None	1963

[a]Data as of 1970, from a survey conducted by the city of Midland, Michigan.
[b]Methoxyclor sprayed in 1970.
[c]Benlate sprayed in 1970.

SOURCE: From *Urbanization and Environment, the Physical Geography of the City* by Thomas R. Detwyler and Melvin G. Marcus. © 1972 by Wadsworth Publishing Company, Inc., Belmont, California 94002. Reprinted by permission of the publisher, Duxbury Press.

was routinely applied to uninfected elms to protect them from the feeding elm bark beetles that carry the spores of the fungus. DDT spraying has now been largely replaced by applications of lindane or methoxychlor. Sanitation, or removal of diseased trees, has also been an extremely effective means of controlling Dutch elm disease (Miller et al. 1969, Neely 1972). A recent development in the control of Dutch elm disease employs the broad-spectrum fungicide benomyl applied to the soil beneath elms (Biehn 1973, Stipes 1973), injected directly into the stems (Gregory et al. 1971, McWain and Gregory 1971), or applied to the foliage (Smalley et al. 1973). Most Dutch elm disease control programs involve insecticide application and sanitation. The use of systemic fungicdes remains largely experimental (Smalley et al. 1973).

Recently, increased emphasis has been placed on planting trees that are particularly able to develop in the city (Michalko 1972, New Jersey Federation of Shade Tree Commissions 1965, Walter 1972, American Society of Planning Officials 1968). Table 3-4 shows some tree species with special characteristics making them especially suitable or unsuitable for urban planting.

Despite this recent trend, urban and suburban control programs still rely heavily or entirely on the use of pesticides. The tendency, too, is to treat all of the trees whether or not any pest problem has been found. One exceptional insect control program has been in operation in Berkeley, California, since 1970. The city has virtually eliminated the use of pesticides for that purpose and has reduced expenditures for pesticides by about $22,500 per year (Olkowski 1974, Olkowski et al. 1974), by (a) establishing populations of parasites that control several non-native pests; (b) using various mechanical methods of control, such as high-pressure water sprays, sticky adhesive bands, and selective pruning techniques; and (c) improving monitoring procedures so that fewer unnecessary treatments are made when chemicals are used (Olkowski et al. 1974).

This approach may act as a model for programs in other cities. At present, however, urban trees are generally subject to regular calendar spraying or whatever is felt to be necessary to ensure their survival. This situation is in large part due to the lack of input of professional, unbiased, and biologically sound advice in the development of control programs. The integrated control program in Berkeley was developed through a cooperative effort involving the city's Recreation and Parks Department and members of the Department of Entomological Sciences,

TABLE 3-4 Trees with Special Characteristics for Urban Planting[a]

	Characteristics	Kinds of Trees
↑	Resistant to extreme city conditions	Tree of heaven, box-elder, white mulberry (all grow where nothing else will, but are not generally desirable)
	Resistant to most city conditions	Norway maple, horse chestnut, green ash, hawthorn, ginkgo, London plane, honey locust, red oak, linden, European hornbeam
	Fast growth	Red maple, white ash, ginkgo, green ash, honey locust, European larch, cucumber tree, pin oak, black locust, mountain ash, American linden, Chinese elm, pitch pine, red pine, white pine, Scotch pine
	Privacy and protection	Evergreens or mixtures of evergreens and deciduous trees, euonymus, firethorn, lilacs, mock orange
Favorable	Resistant to air pollution	Birch, catalpa, elms, ginkgo, hawthorns, London plane, magnolia, tree of heaven, English oak, tulip tree
	Resistant to most biotic disease and insects	Russian olive, ginkgo, honey locust, Kentucky coffee tree, golden rain, sweet gum, cucumber, sweet bay, magnolia, sour gum, sourwood, and cork
	Resistant to ice damage	Beech, catalpa, ginkgo, golden rain, hawthorn, hop hornbeam, horse chestnut, locust, oaks, yellow birch
	Flowers and color	Dogwood, English hawthorn, honey locust, golden rain, tulip, cucumber, sweet bay, magnolia, beach plum, Japanese cherry, black locust, fringe-tree, flowering fruit trees
↓	Sweet scents	Silver wattle (*Acacia*), amur maple, silk tree, English hawthorn, laurel or sweet bay, sorrel or sourwood,

TABLE 3-4 (Continued)

	Characteristics	Kinds of Trees
Favorable		magnolias, citruses, pines and balsams
Favorable	Edible fruits that attract birds	Cotoneasters, flowering crabapple, sapphire berry, spreading juniper, hawthorn, sumacs, hollies
Detrimental	Weak-wooded (break in storms)	Chinese elm, silver maple, mountain ash
Detrimental	Prone to drip on cars	Birches, elms, lindens
Detrimental	Prone to clog drains and sewers	Willow, poplar, silver maple, catalpa, elms, some locusts, lindens, black walnut, horse chestnut

[a] After American Society of Planning Officials 1968

SOURCE: From *Urbanization and Environment, the Physical Geography of the City* by Thomas R. Detwyler and Melvin G. Marcus.

University of California, Berkeley. Alternatively, professionals could be trained as "practitioners of plant welfare"--with broad knowledge of ecology, physiology, entomology, pathology, and soils--to provide the necessary advice and services in any urban or suburban community.

The present tendency to overuse chemicals may also be due in part to the aversion of suburbanites and city-dwellers to the mere presence of pests, especially insects, around their homes. This aversion is too often reinforced by public relations material emanating from the USDA and some state extension services, and supported by the media, stressing the negative effects of the organisms with too little regard for the positive or neutral effects of the majority of them. An important aspect of the Berkeley program is the explanation of the department's pest management decisions to concerned citizens. The department is careful to differentiate between insects that damage trees and those that are simply annoying and to place first priority on the monitoring and control of the former.

Pest control measures in urban and suburban areas are among the most intensive anywhere, because of the stressful conditions under which many of the trees are grown and the high values placed on individual trees. Unfortunately, intensive management in these areas often means a high level of input, frequently with inadequate attention to evaluation and the development of ecologically sound pest management strategies.

REFERENCES

Abrahamson, L. P., and J. D. Harper (1973) Microbial insecticides control forest tent caterpillar in southwestern Alabama. USFS Res. Note. SO 157. p. 3.

American Society of Planning Officials (1968) Trees in the City. Planning Advisory Service Report No. 236. Chicago.

Andresen, J. W. (1973) Forestry thrives in Chicago. Am. For. 79(4):12-15, 58-59.

Berry, F. H., and W. Lautz (1972) Anthracnose of eastern hardwoods. Forest Pest Leaflet No. 133. Washington, D.C.: U.S. Forest Service.

Biehn, W. L. (1973) Long-term protective action of benomyl soil treatment against Dutch elm disease. Plant Dis. Rep. 57:35-57.

Blais, J. R. (1973) Control of spruce budworm: current and future strategies. Bull. Entomol. Soc. Am. 19(4):208-213.

Clement, R. C., and I. C. T. Nisbet (1972) The Suburban Woodland: Trees and Insects in the Human Environment. New York: National Audubon Society Conservation Rep. No. 2.

Dahlsten, D. L. (In press) Bark beetle control. *In* The Lindane Controversy. Berkeley, Calif.: USFS Pacific and Southwest Forest and Range Experiment Station.

Detwyler, T. R. (1972) Vegetation of the city, pp. 229-257. *In* Urbanization and Environment. Belmont, Calif.: Duxbury Press.

Elson, P. F. (1974) Impact of recent economic growth and industrial development on the ecology of northwest Miramichi Atlantic Salmon (*Salmo salar*). J. Fish. Res. Board Can. 31(5):521-543.

Fowells, H. A. (1965) Silvics of Forest Trees of the United States. Agriculture Handbook No. 271. Washington, D.C.: U.S. Department of Agriculture. pp. 373-383.

Gill, D., and P. Bonnett (1973) Nature in the Urban Landscape: A Study of City Ecosystems. Baltimore: York Press.

Gregory, G. F., T. W. Jones and P. McWain (1971) Injection of Benomyl into Elm, Oak and Maple. Northeast For. Exp. Sta. Res. Paper No. NE-232.

Johnson, R. L. (1972) Genetically improved cottonwood--a research and development success. pp. 113-119. *In* Proc. 1972 Nat. Conv. Soc. Am. For.

Kerswill, C. J. (1967) Studies on effects of forest sprayings with insecticides, 1952-1963, on fish and aquatic invertebrates in New Brunswick streams: introduction and summary. J. Fish. Res. Board Can. 24:701-708.

Lewis, C. E. (ed.) (1970) Shade Tree Evaluation. Urbana, Ill: International Shade Tree Conference, Inc.

Long, A. J., H. D. Gerhold, and M. E. Demeritt, Jr. (1973) Metropolitan Tree Planters Survey: Initial Results. Res. Paper No. 41, University Park: Pennsylvania State University, School of Forest Resources.

Massey, C. L., and N. D. Wygant. (1954) Biology and Control of the Engelmann Spruce Beetle in Colorado. USDA Forest Service Circ. 944.

McConnell, R. (1973) Successful tree planting in older urban streets. The Shade Tree (N.J.) 46:66-69.

McWain, P., and C. E. Gregory (1971) Solubilization of Benomyl for Xylem Injection in Vascular Wilt Disease Control. Northeast For. Exp. Sta. Res. Paper NE-234.

Michalko, J. G. (1972) Unusual trees for street planting. The Shade Tree (N.J.) 45:74-77.

Miller, H. C., S. B. Silverborg, and R. J. Campana (1969) Dutch elm disease: Relation of spread and intensification to control by sanitation in Syracuse, New York. Plant Dis. Rep. 53:551-555.

Miller, P. C. (1970) Age-distribution of spruce and fir in beetle killed forests on the White River Plateau, Colorado. Am. Midland Nat. 83:206-212.

Neely, D., E. B. Himelick, and W. R. Cowley, Jr. (1970) Fertilization of established trees: A report of field studies. Ill. Nat. Hist. Surv. Bull. 30:235-266.

Neely, D. (1972) Municipal control of Dutch elm disease in Illinois cities. Plant Dis. Rep. 56:460-462.

New Jersey Federation of Shade Tree Commissions (1965) Trees for New Jersey streets. New Brunswick, N.J.: Rutgers University.

Odum, E. P. (1971) Ecological principles and the urban forest, pp. 78-81. *In* The Rolc of Trees in the South's Urban Environment, U.S. Forest Service Symposium, Jan. 31 - Feb. 3, 1971. Athens: University of Georgia.

Olkowski, W. C. (1973) A Model Ecosystem Management Program. Proceedings of the Tall Timbers Conference on

Ecological Animal Control by Habitat Management, Vol. 5, pp. 103-118.
Olkowski, W., C. Pinnock, W. Toney, G. Mosher, W. Neasbitt, R. Van den Bosch, and H. Olkowski (1974) An Integrated Insect Control Program for Street Trees. Calif. Agr.
President's Advisory Panel on Timber and the Environment (PAPTE) (1973) Report. Washington, D.C.: U.S. Government Printing Office.
Smalley, E. B., C. J. Meyers, R. N. Johnson, B. C. Fluke, and R. Vieau (1973) Benomyl for practical control of Dutch elm disease. Phytopathology 63:1239-1252.
Smith, D. M. (1969) Adapting forestry to megalopolitan southern New England. J. For. 67:372-377.
Smith, W. H. (1971) Technical review: Trees in the city. J. Am. Inst. Plan. 36:429-436.
Stipes, R. J. (1973) Control of Dutch elm disease in artifically-inoculated American elms with soil-injected benomyl, captan, and thiabendazole. Phytopathology 63:735-738.
Struble, G. R. (1973) Biology, Ecology, and Control of the Lodgepole Needle Miner. USDA Forest Service Tech. Bull. 1458.
Telford, A. D. (1961) Lodgepole needle miner parasites: Biological control and insecticides. J. Econ. Entomol. 54(2):347-355.
Toulan, N. (1970) The distribution and value of open land in the Philadelphia area. *In* D. A. Wallace (ed.), Metropolitan Open Space and Natural Process. Philadelphia: University of Pennsylvania Press.
U.S. Department of Agriculture (1972) Extent and Cost of Weed Control with Herbicides and an Evaluation of Important Weeds. Washington, D.C.: Economic Research Service, Extension Service, and Agricultural Research Service.
U.S. Department of Agriculture (1973) Major Uses of Land in the United States, Summary for 1969. Economic Research Service, Agricultural Economic Report No. 247. Washington, D.C.
U.S. Forest Service (1972) USDA Final Environmental Statement on Cooperative 1972 Gypsy Moth Suppression and Regulatory Program. Washington, D.C.: U.S. Department of Agriculture.
U.S. Forest Service (1973) Forest and Wind Barrier Planting and Seeding in the United States. Washington, D.C.
U.S. Forest Service (1974) Forest Tree Seed Orchards: A Directory of Industry, State, and Federal Forest Tree Seed Orchards in the U.S. Washington, D.C.

University of Massachusetts, Cooperative Extension Service (1970) Symposium on Trees and Forests in an Urbanizing Environment. Planning and Resource Development Series No. 17, Amherst.

Waggoner, P. E., and J. D. Ovington (1962) Proceedings of the Lockwood Conference on the Suburban Forest and Ecology. Conn. Agr. Exp. Sta. Bull. No. 652. New Haven.

Wakeley, P. G. (1954) Planting the Southern Pines. Washington, D.C.: U.S. Forest Service, U.S. Dept. Agr., Agr. Monogr. 18. p. 233.

Walter, R. F. (1972) Planting municipal parking areas. The Shade Tree (N.J.) 45:4-7.

Waters, W. E. (1970) The case for forest entomology. J. For. 68(2):73-77.

Wester, H. V. (1968) Spraying and other controls for diseases and insects that attack trees and shrubs. Tree Preservation Bull. No. 6. Washington, D.C.: National Park Service.

Wickman, B. E., and R. L. Lyon (1962) Experimental control of the mountain pine beetle in lodgepole pine with lindane. J. For. 60(6):395-399.

Wilson, L. F., and P. C. Kennedy (1970) Pine root collar weevil hazard zones for red pine in lower Michigan. USFS Res. Note NC-104.

4

THE IMPACT OF PUBLIC POLICY ON FOREST PEST MANAGEMENT

A forest pest has been defined as an organism (excluding man himself) that diminishes the value of a resource in which man is interested. Public policy regarding forest pest control, therefore, must be considered as a part of broader policies on forest management principles and objectives (PAPTE 1973).

The problem of management is greatly compounded with respect to forests in contrast to other crops for four main reasons: (a) There is wide variety in the principal management objectives of forest land, depending on size and type of ownership and site conditions. (b) Forestry, to a greater extent than other branches of agriculture, employs techniques that manipulate natural forces. (c) The long period, often measured in decades, between initiation of a management plan and its maturation means that the plan's objectives may require dramatic change before maturation because of changing economic assumptions and because of the long exposure to natural risks such as fire, insects, and disease. (d) There is seldom a single management objective; rather, the owner (including the federal government) must respond to a wide variety of often conflicting demands that are themselves in flux.

To a large extent, these factors reflect the relatively low value per unit area (in any single year) of forest compared to other crops; but they also reflect the fact that until recently the emphasis in American forestry was, first, on minimizing losses of commercial value in old-growth stands, and later, on harvesting and regeneration rather than on management of young and intermediate-aged stands.

Commercial forest management in the United States is now in transition, moving toward intensive management that pays close attention to inventory and to the cost and effectiveness of particular management practices. Intensive commercial forest management focuses on smaller tracts, with the explicit objective of frequent manipulation of the forest environment to maximize the yield of wood products that exhibit desired characteristics.

We have isolated the following elements that bear on forest management generally and that will surely become more explicit and systematized as the level of forest management intensifies:

1. Historic patterns of land ownership and use that set biological and economic limits on the objectives a forester may realistically elect to pursue
2. The present demand for the forest's "product"
3. Anticipated demand at the time of future harvesting
4. The direct cost of a particular management practice
5. The forest manager's knowledge of the options open to him for achieving a particular objective
6. Public health and environmental restraints limiting options that might otherwise be available
7. Political demands on the managing agency in those instances where public agencies play a significant role in management decisions.

With the exception of restraints derived from historic patterns of land ownership and use, public policy may affect each of these ingredients of forest management. And as forest management becomes more intensive, public policies will have an increasingly important effect on decision making.

Like other major areas of public attention, forest management is strongly influenced by forces outside the forest industry. Federal energy policy, for example, could have an important but qualitatively uncertain effect on commercial forest management. National policy on energy affects forest policy at the supply end by an influence on available harvesting techniques and at the demand end by an impact on manufacturing processes and on the cost of wood products in competition with substitute materials. For example, it has been estimated that for comparable purposes in housing construction, aluminum and concrete blocks consume eight times as much energy as wood in their manufacture (USFS 1972c); national fuels and electrical energy policy may therefore have a profound effect on forest management by influencing both current demand and future demand.

Because of such uncertainties, the present discussion is limited to those public policies (a) that have a clearly discernible or potentially decisive influence on the pest control component of forest management decision making and (b) that are susceptible to administrative or legislative change. The first of these criteria eliminates from our present attention the impact of national energy policy; the second eliminates such important factors as historic patterns of land ownership and use.

FEDERAL POLICIES

Of the nearly 500 million acres of commercial forest land in the United States (USFS 1972a) a little more than one-fifth is owned by the federal government. The Multiple-Use Sustained Yield Act of 1960 directs that such lands be administered for "outdoor recreation, range, timber, watershed, and wildlife and fish purposes." These objectives are superimposed on the objectives derived from the original Enabling Act of 1897, to "improve and protect the forest . . . for the purpose of securing favorable conditions of water flows, and to furnish a continuous supply of timber for the use and necessities of the citizens of the United States" (Title 16, USC Sections 472, 528). The impact of federal agencies, however, principally the USFS, is not restricted to federally owned forest lands. The USFS is the major research organization with respect to forest practices. In the western states, especially, where federal forest land makes up more than half the total, the USFS has an important influence on the management practices of all forest land, directly through its research efforts and its administrative responsibility for federal cost-sharing programs, and indirectly through its influence on the theory of forest practices applied to surrounding lands and through its effect on the forest products economy of the region.

The Secretary of Agriculture, through the USFS, administers the Forest Pest Control Act of 1947 (16 USC Sections 594 et seq.). The language of the Act reflects the post-war enthusiasm for direct, usually chemical, means of pest control. Its purposes are set forth in explicit recognition of the economic importance of forest industries, particularly in the western states, where the management of federally owned forests has a strong influence on local economies.

There are two branches to the operative portions of the Act. First, the Secretary of Agriculture is directed to conduct surveys of and control programs against "ravages of bark beetles, defoliators, blights, wilts, and other destructive forest insect pests and disease. . . ."

Second, in recognition of the checkerboard pattern of public and private ownership of much timber land, the Act provides for cooperative pest control programs on private land, subsidized by cost sharing with federal funds.

The cost-sharing provisions of the 1947 Act are an important policy factor affecting forest pest management practices. The cost of chemical control is prohibitive on small privately owned tracts surrounded by public forest land unless a substantial subsidy is available.

On federally owned commercial forest land, another important limitation on the freedom of forest managers to opt for choices otherwise available is the requirement of the Knudsen-Vandenberg Act of 1930 that a percentage of revenue from timber harvested be applied to regeneration and other cultural practices *on the same area* from which the timber was cut. As noted earlier, most forest pest control on commercial forest land is silvicultural. The area-dependent requirement for allocating Knudsen-Vandenberg funds tends to limit the choice of management strategy, because it does not permit diversion of such funds to areas where early silvicultural application may avert the need for a crisis-management approach at a later time.

A third federal factor is the historic role played by the USFS as economic guardian of local economies in the West. This role is reflected in the 1947 Forest Pest Control Act's declaration of policy, providing that one of the Act's purposes is to reduce the "ravages" of forest pests in order to "promote the stability of forest using industries and employment associated therewith . . ." (16 USC Section 594-1).

Other more recently adopted federal policies influence forest pest management by restricting the available management options in the interests of environmental or public health concerns. Chief among these is the registration requirement for chemicals by the EPA, under the Federal Environmental Pesticide Control Act (FEPCA) of 1972 (PL 92-516). FEPCA will have a major impact on the kinds and scope of research that will be carried out on pest control techniques. Section 5 of that law, for example, will require rather extensive information for the granting of experimental use permits for new chemicals or other technologies. If the information required under the *final regulations* is very extensive, this section may well limit the number of "alternative" techniques that pass this preliminary step prior to field testing. Section 7 would require the monitoring of all pesticide-producing establishments, including research laboratories that produce test materials such as various microbial insecticides. Although some limitations undoubtedly should be placed on such establishments, enough flexibility should be built into the regulations that it does not become virtually impossible to develop alternatives to current pesticides.

A second federal policy that has already had a substantial effect on forest pest control is the National Environmental Policy Act of 1969 (NEPA). Through judicial interpretation, the Act's provision for an environmental impact statement has been made applicable to many practices that heretofore were left almost exclusively to the discretion of the pertinent federal agencies, including a wide variety of pest control and cutting programs. The simple exercise of preparing impact statements has had an effect on the agencies involved. Until now, most of these impact statements have been based largely on guesswork--and they have been severely criticized for that reason (see the tussock moth case study, p. 108). To reduce such uncertainties, and also to satisfy the provision that inaction be regarded as one of the alternatives, agencies like the USFS will have to expand their environmental research effort considerably.

STATE POLICIES

Every state has some legislation regulating private forest practices, but only 15 states have true forest practice acts that go beyond the historically custodial approach that stressed fire control and eradication of severe pest epidemics (Title 16, USC Sections 475, 528). None of the acts on its face reflects emerging concepts of integrating pest control with other forest management practices; most emphasize stocking requirements following cutting, and several require cutting plans enforced by permit conditions or provisions for state-funded restoration, reimbursed by liens on the affected properties. Several, however, including the recently enacted Oregon and California acts, are amenable to an integrated approach through rule making by district forest practice boards.

The California act, effective January 1, 1974 (but not to be fully implemented until regional rule making is completed in the fall of 1974) contains strong emphasis on environmental protection, particularly water quality and erosion control. The role of pest control in management plans required by the act is presently unclear, although the issue appears to be recognized by regional committees now drawing up rules for adoption by the statewide board.

The Oregon act, effective July 1, 1972, upon which California's was modeled, has been followed by specific rules dealing with chemical control of forest pests. The focus of these rules is on maintenance of water quality, however, rather than on integrated forest pest management. The enforcement mechanism involves prior notice by an operator to the state forester of significant activities specified in the act or implementing rules. It is reported that most such notices since the effective date of the act have concerned logging or road-building operations. Control of weed brush and trees along roadways is, however, widespread in Oregon, but this practice is not within the notice requirements of the act.

Thus it is difficult to evaluate the impact of state legislation on forest pest management. About all that can be said with confidence is that the emphasis on water quality and the general sensitivity to chemical pest control methods reflected in the most recent state forest practice legislation suggest increasing restrictions on chemical control practices. Increased attention to the integration of pest control with other silvicultural practices seems the most likely consequence.

CASE HISTORIES

To illustrate the effect of public policies on forest pest management, we have selected four case histories of particular forest pest problems for detailed consideration. The cases, discussed below, are the cycle of white pine blister rust control; the recent Douglas fir tussock moth infestation; the gypsy moth in the Northeast; and the recent southern pine beetle outbreak.

The Cycle of White Pine Blister Rust Control

The history of control efforts aimed at blister rust provides an apt illustration of the role of public policy in forest pest management. Caused by an exotic fungus, the disease's course and the pattern of initial public response are unusually well charted. The disease affects several of the most valuable of timber species; thus the question of present and future economic demand for the host was properly taken for granted in the early election to pursue vigorous control measures. Second, both the value of white pine and the timing of the introduction of the disease in the history of white pine forestry focused early attention on the disease as a factor in second-growth management--as distinguished, for example, from insect and disease pests that have chiefly been a threat to old-growth inventory and thus not readily susceptible to silvicultural manipulation. At various times, both chemical and sanitation measures have been attempted, as well as physical eradication of the alternate host. Finally, control theory has now evolved to a fairly sophisticated level that takes primary account of the cost of control in light of expected return.

Initially, the problem of control was approached almost exclusively as a question of forest pathology in isolation from the broader context of forest management. After a spectacular spread in white pine stands of the Adirondacks in 1919, prompted by favorable weather conditions, a leading forest pathologist wondered "whether this problem will become a material factor in forest regulation, or whether it is after all a more or less isolated problem in forest pathology" (Snell 1928).

White pine blister rust is caused by *Cronartium ribicola*, which is an introduced, virulent rust fungus that attacks five-needle pines, notably eastern and western white pines and sugar pine. It was first detected at Geneva, New York, in 1906, and within a decade it had gained the status of a major forest pest in the Northeast. A separate infestation originated at Vancouver, British Columbia, in 1910. A decade later it had spread to Idaho, and by the late 1930s the disease had become established on favorable sites in the Pacific Northwest and California.

After reviewing preliminary findings from plots established in response to the 1919 epidemic, indicating 50 to 100 percent mortality in young growth, but decreasing susceptibility in older trees, Snell (1928) concluded that "blister rust is a serious menace to white pine in the Adirondacks. . . ." He recommended a massive eradication program.

The life cycle of the fungus requires an alternate host of the species *Ribes* (e.g., currants and gooseberries). The infecting spores are windborne from the alternate host but seldom remain viable for more than a few hundred feet after their release. The spores travel and germinate best under moist, cool conditions.

The spores penetrate the needle fascicle, and the fungal mycelia eventually girdle the cambium. The disease is readily detected in the field by "flags" of dead needles on infected branches and characteristic cankers on the bark at more advanced stages. The cankers contain the fruiting bodies for spores that reinfect *Ribes*, completing the two-host cycle.

Control of the disease involves suppression of the alternate host; the most widely used technique is mechanical removal of the currant or gooseberry bushes in control areas; eradication using chemical herbicides has been attempted, with inconclusive results.

Ribes eradication was begun on a small scale in the northeastern region in 1918. By 1921, the USDA had entered into cooperative agreements with practically all the western states to guard against spread of the disease and to provide for immediate *Ribes* eradication efforts (Detwiller 1921). Experimental eradication was undertaken in western white pine stands in Idaho in 1924-1925, and by 1934 concerted control efforts were under way throughout the Pacific Coast region.

Throughout the Depression "blister rust crews" were a major part of the Civilian Conservation Corps. Indeed, it is impossible now to segregate from the massive blister rust control efforts undertaken throughout the 1930s the respective influences of genuine attention to forest pathology and the independent role of federal agencies in furnishing employment as a part of New Deal reconstruction. In any event, by 1944, a USFS pest control official (Martin 1944) could write confidently that:

> *Ribes* eradication is effectively controlling the rust on extensive areas in the important white pine regions of the United States. . . . Although blister rust has not been reduced to such a degree in those portions on which the *Ribes* have been suppressed, further damage to the pine stands has been insignificant. On these protected areas control of blister rust has been established.

Such optimism, however, was premature. Not only did the goal of total eradication continue to elude control efforts, but the cost of mechanical *Ribes* eradication and the need for follow-up treatment promised an endless and expensive program. By 1950, 1,825,000 acres had been mapped for intensive *Ribes* removal in the western states; two-thirds of this was federal land. Of this total, nearly a million acres had received a first treatment, nearly half a million acres had been treated a second time and 200 thousand acres were under "maintenance" treatment involving systematic resurvey and eradication efforts. Most of this acreage was in old-growth stands in the Sierra Nevada and the Cascade and Rocky Mountains. The cost of the control efforts in California and Oregon alone exceeded $1.5 million in 1947 and averaged close to $1 million per year for the decade from 1945 to 1955.

The criteria for selecting control areas placed major stress on the density of *Ribes* and the site susceptibility to occurrence of the rust, apart from the economic significance of white pine on such sites. It soon began to appear, however, as logging of old-growth mixed conifer stands accelerated in the housing boom during and after the Korean war, that *Ribes* spp. were prominent successional species in logged-over areas. It followed that because young-growth white pine was highly susceptible to the disease, *Ribes* eradication would have to be a major expense in forest management if these pines were to be maintained in the mixture of second-growth conifers after the removal of old growth.

By the 1950s the early criteria for selecting eradication zones had come into serious question. The difficulty of eradication after fire, logging, and other disruptions raised in the USFS "the fundamental question of the law of diminishing returns. . . . It is therefore important to assess [logged over] lands very carefully as to the amount of pine present after logging before investments are made in control work." Again, "instead of control units encompassing entire drainages, a high degree of selectivity [was suggested], restricting control work to areas of good site and stockings" (Benedict 1950).

It was not long before a theoretical economic framework emerged for systematically carrying out this flexible and selective approach in California and Oregon. In 1954, Vaux showed that the return on a given investment in sugar pine management could be nearly doubled if substantial funds restricted to *Ribes* eradication were diverted to other management techniques, including road building and sanitation cutting, coupled with selective *Ribes* eradication on small tracts with a high relative stocking of sugar pine and high growth potential (Vaux 1954). Vaux criticized past practice:

> Work on such areas has contributed substantially toward the objective of *disease control*. But some of it has contributed relatively little toward the more meaningful public goal of *growing more sugar pine*. . . . [I] is important to distinguish the means from the ends. (Emphasis in original)

The break with past policy, signaled by Vaux's important and widely disseminated thesis, led eventually to abandonment of *Ribes* eradication as a major federal policy.

By 1969, *Ribes* eradication had been discontinued in California and most of the Pacific region except in intensively managed sugar pine plantations (California Forest Pest Control Action Council 1969). In the Pacific Northwest, "*Ribes* spp. eradication, as a method of white pine blister rust control, has been completely phased out . . . following a final [U.S. Forest Service] evaluation completed in the Spring of 1962" (USFS 1973). In California, remaining efforts at direct control in the southern Sierra were discontinued in 1971 "when results of scouting indicated that the disease is intensifying too rapidly in this zone to be contained by eradication of infection centers." (USFS 1972b). In the northern Rockies, despite white pine mortality approaching 90 percent in some areas, no significant control effort has been undertaken for several years. *Ribes* eradication continued in the Lake states, northeast and southern Appalachia during 1971 and 1972, on areas managed for commercial white pine regeneration, and in parklands; total acreage treated each year, however, probably did not exceed 1,000 acres.

Of the seven elements of public policy listed earlier, the present and anticipated demand for white pine lumber and, perhaps, limitation on the USFS's knowledge of options open to it are the most important. White pine, because of its clear, easily worked grain, has historically brought a high price for construction and millwork. It enjoys a special place in the hierarchy of commercial softwoods. The decision to launch a major control effort at a pathogen that clearly threatened its survival as an important commercial species thus came naturally and was not seriously questioned until an aggravated disparity appeared between costs and benefits.

The lesson of the blister rust experience is that pest control that focuses sharply on the pest tends to overlook the fact that pest control should be merely one of several practices within an overall management scheme. The historical practice of singling out pest control as a separate activity in forest management tends to precipitate a bureaucratic momentum that can, and does, lose sight of the importance of weighing pest control decisions against other management options. In short, the long-range benefits tend to be exaggerated while the long-range costs are minimized or overlooked until the disparity becomes embarrassing.

The Recent Tussock Moth Infestation

The recent controversy over use of DDT for controlling a major outbreak of the Douglas fir tussock moth in the Pacific Northwest sharply illustrates changes in forest pest control policy and the continuing tension between management objectives that focus on pest control and a broader approach that places control in perspective alongside other management objectives.

The tussock moth (*Orgyia pseudotsugata*) is an indigenous forest pest in the Pacific Coast region. It tends to spectacular outbreaks, causing widespread defoliation of Douglas fir and grand fir in Oregon and Washington and white fir in California, all of which are important commercial species. Five important outbreaks have occurred in California and the Pacific Northwest since 1935, at roughly 10-year intervals (Wickman et al. 1973). The present controversy turns on whether DDT should be used to control the current outbreak, as it was used without much controversy on four outbreaks since 1945.

The caterpillar has a voracious appetite for new foliage. An infestation can cause major defoliation over large areas of timber, with consequent death or pronounced growth loss in the affected trees. A peculiarity of this insect that causes concern beyond its direct economic impact on commercial values is the presence of small hairs shed from the caterpillars that cause an uncomfortable allergic reaction in humans; this characteristic hampers salvage logging in heavily damaged stands and has an obvious impact on recreational values, apart from the aesthetic losses resulting from defoliation.

There appear to be a number of ecological factors that eventually cause natural collapse of tussock moth epidemics. Not all are understood, but one important and well-documented control is a nucleopolyhedrosis virus that infects the larvae. A major cause of the collapse of past epidemics has been the virus, but some infestations appear to die out for other reasons.

The most recent outbreak was first detected at scattered locations in the Wenatchee and Okanogan national forests in the summer of 1971. By fall of that year, larval populations on the Wenatchee National Forest had died from natural causes, while surveys on the Okanogan forest, where the 1971 outbreaks were smaller and less severe, suggested that these populations would increase in 1972 (USFS 1972b).

During the summer of 1972 most of the moth-killed timber on the Wenatchee forest was salvaged. Simultaneously, the infestation on the Okanogan forest appeared over a larger area and new outbreak centers began appearing at other locations in the region. By July, tussock moth defoliation was visible in aerial surveys in 33 townships in Oregon and 10 in southeastern Washington (USFS 1972d). During the summer, the USFS, in cooperation with the Oregon Department of Forestry and the Washington State Department of Natural Resources, conducted experimental aerial applications of two nonpersistent chemical pesticides in areas where the virus incidence was low; the results were inconclusive.

The Forest Service *Fact Sheet* summarizing the 1972 outbreaks contained no estimates of timber volume affected; rather, the statistical analysis stressed acreage in which "defoliation is visible from the air" and gave no data on the severity of mortality, top-kill, or defoliation on the acreage mapped; such estimates became available in 1973. The controversy over the use of DDT had become a major political issue in the Pacific Northwest, and damage resulted in a net loss of about 30 million board feet after salvage logging of accessible timber.

In anticipation of increased losses, the USFS on March 20, 1973--and later the states of Oregon and Washington, several local municipalities heavily dependent on wood products industries, and one major private forest land holder--applied to the EPA for an emergency exemption of DDT under Section 18 of FIFRA [7 USC Section 136a(c)(1)] from that Agency's 1972 deregistration of DDT to permit application of nearly 200,000 pounds of the pesticide to approximately 500,000 acres. The EPA Administrator denied the application on the grounds that, in light of evidence suggesting that a virus-caused collapse of the epidemic would occur during 1973, the economic benefits of control were outweighed by the known environmental risks of such massive applications of this persistent insecticide (U.S. Environmental Protection Agency 1973).

However, during 1973, not only did mortality appear to be higher than expected on areas defoliated during 1972, but the total area of infestation increased to approximately 650,000 acres in Oregon and Washington, with additional infestations occurring in Idaho, Montana, and California. The USFS estimates that 600 million board feet of merchantable timber was killed in 1973, of which a little more than 500 million could be salvaged.

The failure of the anticipated virus-caused collapse to halt the infestation and the magnitude of the summer 1973 losses generated intense pressures to force EPA reversal of its position prohibiting the use of DDT. A crisis climate seemed to prevail. An Interagency Douglas Fir Tussock Moth Committee was formed, headed by the USFS. The natural resources agencies of the affected states joined the USFS in demanding exemption of DDT for tussock moth control. An intense lobbying effort was begun by timber industry trade associations. By the fall of 1973, these efforts had succeeded in having reported out of the House Agriculture Committee a bill (HR 10796) directing the Administrator of EPA to approve promptly any registration application by the Secretary of Agriculture for the use of DDT on forest or agricultural land.

The theme struck by proponents of DDT was that it was "the only chemical which has proven effective against outbreaks of Douglas fir tussock moth in the past" (Ciesla 1973). The proof consisted of convincing evidence that several major outbreaks since 1947 had collapsed following DDT application. However, the causal connection could not be established: not only was adequate statistical comparison with untreated areas lacking, but in all cases the collapse following treatment occurred when collapse from the virus would typically have occurred.

Scientific opinion was sharply divided, as is often the case when research results are insufficient and the issue is controversial. Indeed, industry advocates of DDT blamed the EPA denial of the 1972 application to use DDT in part on a "*Silent Spring* mentality" fostered by a "politics of environmentalism" (Anonymous 1973); research findings that cast doubt on the efficacy of DDT were blamed for the lack of a "scientific consensus" in support of DDT (Fredsall 1973).

On January 3, 1974, the USFS, again joined by the states of Oregon and Washington, renewed its request for emergency exemption of DDT on a contingency basis. The application contended that if normal review time elapsed for consideration of the application by EPA, any approval would come too late to permit the necessary contracts to be let or for personnel to be adequately trained. The application was for DDT use contingent on need as demonstrated by surveys of the 1974 egg hatch to be made later in the spring. The USFS estimated that about half a million pounds of DDT would be applied over as much as 650,000 acres of federal, state, and private forest land.

The EPA was already under considerable criticism for its denial of the similar request in 1973, which was followed by an increase in the acreage of infested timber. The House Agriculture Committee's bill, HR 10796, was widely believed to be aimed at the current infestation, but the larger effect of its passage would have been to strip the EPA of any real control over DDT.

These political considerations surely played some part in the EPA's decision on February 26 to grant the application (Fed. Reg. 1974). As EPA administrator Russell E. Train stressed in a statement announcing the decision: ". . . [T]he efforts to date on the part of the U.S. Forest Service to develop real alternatives to DDT for the control of the tussock moth have been almost totally inadequate--to the point of dereliction." But, "we were left with no realistic, practical alternative . . . ."

After reviewing the alternatives to DDT, the EPA found "uncertainties . . . in the following areas:

> (1) The relationship between the intensity of larval populations and damage to trees;
> (2) The efficacy of controls to prevent damage;
> (3) The exact economic and social impact of a decision not to control the infestation;
> (4) The extent of the virus population this year and its relationship to the potential collapse of some or all the infestations.

Spraying began the second week of June on the Colville Indian Reservation, Washington. By mid-June approximately 15,000 acres had been treated at various locations in Oregon, Washington, and Idaho. The USFS estimated at the beginning of spraying that something over 400,000 acres would be treated by the time spraying was completed in July.

An important condition imposed by the EPA in its order granting the emergency application concerned research requirements aimed at developing registerable alternatives to DDT, the results to be submitted to EPA by the end of 1974.

The tussock moth controversy in the Pacific Northwest has clearly precipitated an important test of whether the environmental awakening of the 1960s, institutionalized in NEPA, FEPCA, OSHA, and other recent legislation, can withstand concerted efforts to erode these advances. The 1972 EPA ban on virtually all uses of DDT represents a national policy intended to protect national, even global, interests. The central policy question, then, is whether sharply focused local interests are sufficiently important to reverse such a decision.

It is useful in approaching this question to test the DDT issue against the array of criteria underlying the understanding of tussock moth population management.

1. Historic patterns of land use: little is known about the conditions that precipitate tussock moth outbreaks, but land use patterns do not now seem significant. The affected species in Oregon and Washington are attacked in both young- and old-growth stands indiscriminately.

2. The current and anticipated commercial demand for affected tree species will probably continue at a high level. Thus, it is not necessary to speculate on the need for protection and control. The difficult part of such analysis is to evaluate the net loss as a basis for systematically weighing the benefits of control, whether by chemical or other means. There are at least three factors to consider in this analysis. One factor is the actual loss from dead timber that cannot be salvaged and would not otherwise be lost to age, fire, windthrow, and other insects and diseases; inaccessibility, shortages of labor and equipment for large-scale salvage over a wide area, and disadvantageous economic conditions may prevent optimum salvage of killed timber. A second factor is the economic impact of growth retardation from defoliated areas without mortality. In the context of the tussock moth controversy, David A. Graham, USFS Branch Chief for Insect and Disease Control in the Division of Timber Management, for the region encompassing Oregon and Washington, stressed that (Graham and Wright 1973)

> we do *not* have an adequate system for measuring, evaluating, or predicting insect- and disease-caused impacts. . . . Basically, we do not have a clear understanding of the concepts and practical implication of pest impacts in the total space-time frame of the resource management process. [Original emphasis]

A third factor is the cost occasioned by disruption of formal management plans and consequent discontinuity in future timber supplies. Without some theoretical framework for measuring these factors, it is not possible to state with confidence what current values are lost, from a local or national viewpoint, by failure to control an epidemic such as the current tussock moth outbreak.

3. The dollar cost of DDT application on the scale envisioned for the current tussock moth infestations is considerable. On federal land the cost is borne entirely by the general public; on state, municipal, and private lands half would be borne by the nonfederal owner under provisions of the Forest Pest Control Act of 1947 (USFS Manual, section 3412.14). The pertinent regulations are

drawn from the perspective of direct insect and disease suppression and do not lend themselves to comparable cost sharing for cultural practices. It appears that at least some of the pressure for the EPA to approve DDT stems from a realization that chemical control would not be undertaken by private timber owners in the absence of the cost-sharing subsidy. Considerable resentment has been expressed in industry circles, for example, over the expense incurred in a cooperative application of Zectran (double application) over approximately 70,000 acres of mostly private land, following the 1973 EPA denial of applications to use DDT. Many of the landowners were dissatisfied with the ineffectiveness of the experimental spraying coupled with the cost of $3.75 per acre, which was higher than anticipated (Larsen 1973). This feeling in the industry is reflected in a resolution adopted by the Forest Pest Committee of the Western Forestry and Conservation Association on December 5, 1973, calling for federal compensation for "all costs, losses, and damages which occurred after the suppression of the epidemic was prevented by unresponsive government."

4. The forest manager's knowledge of control options presupposes sufficient time to implement known measures. It is sufficient at this point to stress again that research conducted in a crisis setting is likely to be unreliable.

The current controversy sketched above illustrates that, among the elements identified earlier as bearing on forest management decision making, considerable tension can develop between public health and environmental constraints and political demands on a managing agency that has a significant role in overall forest management. Where another of our seven elements--"the forest manager's knowledge of the options open to him"--is deficient, environmental and public health restraints give way to the political demands that are naturally generated when the pest is perceived as a serious threat to the local economy.

In its order of February 26, the EPA repeatedly suggested that it was flying blind. The "uncertainties" noted earlier cover virtually every element necessary to sound pest control decisions. Unfortunately, until forest management generally reaches a level of intensiveness that now exists only on a few of the most highly productive sites, important pest control decisions probably will continue to be founded largely on "uncertainties."

The tussock moth has received a considerable, perhaps disproportionate, amount of attention because it is currently a serious problem. This attention has resulted in the insect's inclusion in the expanded research effort of the USDA. Current (1974) plans call for a 3-year research effort and an additional expenditure of $7,890,000 to develop a means of coping with the problem. The research has numerous specific objectives, but the major expectation is that an integrated program of control will be forthcoming.

The Gypsy Moth: A Suburban Insect Pest

Probably no forest insect has been subjected to more public attention than the gypsy moth (*Porthetria dispar*). Aside from the western budworm, none has been treated with insecticides on so large an acreage in the United States. In 1869 this Eurasian insect was introduced into Medford, Massachusetts, as part of a silk production experiment. Several moths escaped, and 20 years later the population exploded. Today the species has achieved pest status in much of New England, New Jersey, and Pennsylvania and is still extending its range (Figure 4-1). Although primarily a northeastern pest, isolated specimens have been seen as far west as Berkeley, California. Millions of dollars have been spent to eradicate and, more recently,

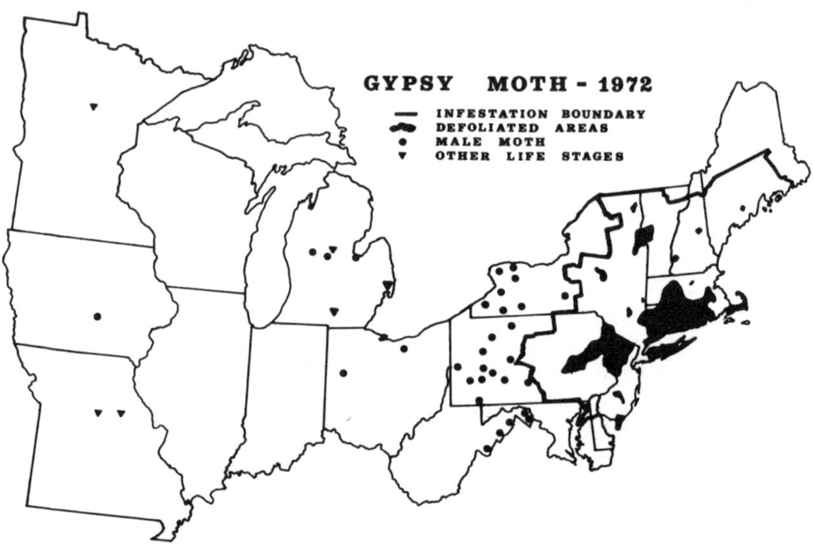

FIGURE 4-1. Gypsy moth distribution in the northeastern states in 1972. (Source: USFS 1973.)

to control the moth, which is now recognized as part of the natural forest ecosystem of the Northeast. In 1971, 1,439,770 acres were defoliated, and 371,900 acres were sprayed with insecticides in New York, New Jersey, and Pennsylvania (Table 4-1). In 1972, 1,369,130 acres were defoliated and 174,180 acres were treated. The primary insecticides used in 1971 were carbaryl (Sevin) and, in test programs, tricholorofon (3,000 acres) and *Bacillus thuringiensis* (1,800 acres) (USFS 1973).

Life History

During the early summer, the flightless females deposit 100 to 600 eggs in a single mass that is attached to the bark of trees or to stones or debris. Hatching occurs the following year in April and May, and after 5-6 weeks the caterpillars pupate and a new adult emerges in mid-July. The defoliating period occurs during late spring and early summer, when the caterpillars are most active. All life stages of the moth are subject to killing at low temperatures, and numerous parasites and predators may greatly reduce the population. Ichneumonid flies, *Calosoma* beetles, white-footed deer mice, and shrews, as well as many birds, prey on this insect. In addition, a wilt disease carried by a polyhedral virus can cause significant mortality. A more complete description of the life history is found in Campbell (1967).

It has been found that the gypsy moth has an *innocuous phase* and an *epidemic phase* (Campbell 1973). During the innocuous phase the population numbers fluctuate for a number of years at a relatively low level; no more than a few hundred egg masses can be found per acre, and defoliation is insignificant. In contrast, the outbreak or epidemic phase involves thousands of egg masses per acre and may result in heavy defoliation of susceptible trees. During this phase, outbreaks reoccur at irregular intervals, usually lasting 2 to 4 years. However, in Massachusetts the outbreak phase has been reported to be 10 to 20 years in duration and has involved repeated defoliations (Clement and Nisbet 1972). The spread of the moth apparently occurs during the early larval stage, when the tiny caterpillars spin threads that enable them to be carried considerable distances by the wind (McManus 1973).

TABLE 4-1 Acreage Defoliated by the Gypsy Moth and Treated with Insecticides, 1971 and 1972

State	Acres Defoliated 1971	Acres Defoliated 1972	Acres Treated 1971	Acres Treated 1972
Connecticut	655,100	513,880	47,000	--
Maine	820	40	--	--
Massachusetts	18,800	20,480	--	--
New Hampshire	3,300	200	--	--
New Jersey	180,600	226,140	55,000	47,412
New York	479,150	177,605	241,000	43,000
Pennsylvania	89,000	404,060	23,000	24,800
Rhode Island	8,500	22,150	5,900	58,968
Vermont	4,500	4,215	--	--
TOTAL ACREAGE	1,439,770	1,369,130	371,900	174,180

SOURCE: USFS 1973.

Effects on the Forest

The impact of the gypsy moth on forested areas has been studied considerably. Egg mass densities of 800 per acre, given favorable survival, may result in 25 percent defoliation; 2,000 egg masses per acre result in 50 to 75 percent defoliation; and 3,000 or more may result in 90 percent defoliation or more. Campbell (1973) has completed a thorough analysis of egg mass data and now has a model available for forecasting egg mass densities. Actual mortality resulting from gypsy moth defoliation is highly variable, as the insect does not affect all sites and types uniformly. The most severe defoliation occurs on the higher ridges or upper slope and crest sites, where the impact of dryness is most pronounced. There is also a preference for certain species (Turner 1963), including apple (*Malus* spp.), aspen (*Populus* spp.), beech (*Fagus* spp.), birch (*Betula* spp.), larch (*Latrix* spp.), linden (*Tilia* spp.), oak (*Quercus* spp.), and willow (*Salix* spp.). Species usually untouched are mountain laurel (*Kalmia latifolia*), ash (*Fraxinus* spp.), tulip tree (*Liriodendron tulipifera*), black walnut (*Juglans nigra*), striped maple (*Acer pennsylvanicus*), balsam fir (*Abies balsamea*), cedar (*Juniperus* spp.), butternut (*Juglans cineroea*), sycamore (*Platanus* spp.), locust (*Robinia* spp.), and dogwood (*Cornus* spp.). In southern New England it is not uncommon to find isolated specimens of tulip tree in full foliage within a highly defoliated area.

Studies by Collins (1961) in Connecticut indicate that about 5 percent of the oaks may die as a result of defoliation. However, he believed that this merely represents a compression of 5 years' mortality into one year, as the mortality rate decreases in subsequent years. Collins's findings may not apply generally, as other studies have given much different results. In Massachusetts, Tierney (1947) indicates oak mortality ranging from 10 to 80 percent after 3 years of heavy defoliation. Kegg (1971) showed very severe mortality in a Newark watershed. That severe tree mortality occurs cannot be denied; however, deciduous species usually recover from a single defoliation. In contrast, evergreens, especially hemlock (*Tsuga canadensis*) may be killed by a single defoliation. Deciduous trees generally refoliate, so that by the end of summer the forest is essentially in full foliage again.

Tree mortality is usually greatest on the drier, less productive upper slopes and crests. The probability of death of hardwoods has been low, of pine moderate, and of hemlock relatively high. The loss of individual trees, either planted or natural, around suburban homes can occur.

Control

The fact that epidemic outbreaks occur within urban-suburban environments where large numbers of people live and engage in recreation has generated much controversy at the local and state levels over control techniques. Aerial spraying has been the standard procedure for many years, and for a long time it was thought that eradication of the insect might be possible (Doane and Schaefer 1971). Massachusetts has the longest record of active programs against the pest and New York State also has had a large program for many years. In New York an attempt was made to hold the line on the expansion of gypsy moth range. While each state organization has varied somewhat in the approach to the problem, most now look at this insect as being a part of the natural complex. New Jersey and Pennsylvania at first attempted to eradicate the insect but gave up this idea several years ago; control is now carried out only where the problem may be especially serious.

Several insecticides have been used against this pest: in New York State some 400,000 lb of lead arsenate was used prior to World War II; from 1945 to 1971 about 3.5

million lb of pesticide has been applied (Reese and Becker 1972). Most of this has been DDT and Sevin (carbaryl), with complete elimination of DDT since 1966. All of the states involved have developed or adopted the idea of integrated control and are utilizing biological, chemical, and cultural methods to combat the pest.

Federal policy was and is involved in the decisions on gypsy moth control. This insect had been a quarantined pest and as such its control was a responsibility of the Plant Pest Control Division of the APHIS (the Animal and Plant Health Inspection Service). Once the insect was accepted as a part of the natural complex it was no longer quarantined within its range. A few years ago an agreement was made between APHIS and the USFS that essentially gives the Forest Service responsibility in areas where the moth is already established. The general responsibility for gypsy moth work now rests with two federal agencies, several states, and individual landowners. Policies vary but are well-coordinated in view of the wide variety of groups involved.

The idea of pest management is not new, but the principles are being applied more forcefully than in the past. The system makes sense, and people are beginning to accept these ideas in preference to a reliance on one chemical. Waters has described the basic premises in several of his papers (1971, 1973a, 1973b).

Currently, there are a number of procedures available for control of the gypsy moth. Carbaryl is the most widely used chemical, though it has undesirable side effects, especially on nontarget insects. The material *Bacillus thuringiensis* has been tested and is used in a limited way in control (Dunbar et al. 1973, Lewis and Connola 1966). Biological systems have been investigated, and several states have rearing projects in progress. Large numbers of parasites have been reared and released in some of these locations. The early mechanical methods may still have a place in some situations where large acreages are not involved.

Some of the methods used are expensive, and some of the newer methods that may be developed could be more costly. This is an insect of importance to owners of residential properties. Since the value of infested trees in such circumstances may be relatively high (Payne et al. 1973), higher control cost can be tolerated.

The gypsy moth has received considerable attention for most of the past half-century, and it will receive more. There is a large research effort being expended to obtain more complete answers to the problem. It is hoped that this work will lead to an effective pest management plan. The USFS in cooperation with the states and other agencies has conducted an expanded R&D program for several years, with James L. Bean as Program Manager. The Congress has designated the gypsy moth as one of the three insects on which to concentrate. Plans call for an additional 3-year effort to solve pressing problems. The USDA work will be managed by several cooperating agencies of the U.S. Government and the various states. An effective gypsy moth council has assisted in this research effort and will no doubt continue to have a strong positive effect on the programs.

Modeling studies of Campbell (1973) on gypsy moth population data gathered between 1911 and 1931 in eastern New England indicate several significant findings. Once an area has reached a high overall density of moths, a massive outbreak can be expected to continue for many years. Second, almost any woodland environment, regardless of its inherent capacity to maintain sparse gypsy moth populations, will probably be characterized by a high-density resident moth population if it is located in the vicinity of a massive outbreak. These two conclusions suggest "that a massive outbreak by this insect, left alone, may not only persist for many years, but also will probably expand in size." Therefore, spot treatments can be relied upon only to save foliage and keep trees alive. As Campbell indicates, the options for coping with massive outbreaks are limited:

> 1. We can do nothing and hope that one of the following naturally occurring processes will terminate the OUTBREAK phase:
>
> Populations will increase explosively to extremely high from year (n -1) to (n) and then recede abruptly to innocuous levels in year (n + 1).
>
> The area covered by the entire OUTBREAK system will be virtually stripped of its foliage in the summer of year (n + 1), and subsequent density levels will return to the INNOCUOUS phase.
>
> Precipitation will be extremely high in June of year (n + 1) across the entire area.
>
> The mean minimum winter temperature during the coldest month between years (n) and (n + 1) will approach zero °F across the entire area.
>
> 2. Or, we can continue to treat places where the values threatened by the insect are known to exceed both the known direct cost and estimated indirect costs of control, in the knowledge that such treatments will not solve the underlying problem.

> Clearly, what is really needed here is an environmentally acceptable treatment that could be applied *effectively* to the entire massive outbreak, but no such treatment is now available.

Aware that they are dealing with a two-phase system, future researchers will be able to focus on one phase or the other, knowing that the system will move from one phase to the other if the work is initiated in the transition. The model fails to indicate a mechanism by which the innocuous phase is transformed into the outbreak phase except through an increasing diversity level possibly due to overlap between areas of outbreak and innocuous phases. Perhaps more answers will be forthcoming from the research of the next 3 years.

The Southern Pine Beetle: A Native Forest Pest

The southern pine beetle, *Dendroctonus frontalis*, is a destructive pest of pine in the South. Since as early as 1899, research on this insect has been almost continuous (Hopkins 1899), although research activity has been sporadic. This native insect is widespread throughout the pine regions of the South. It ranges from Maryland to Florida, west to Texas, and south through Mexico to Honduras. The insect will attack and kill any species of pine in its range and has been especially destructive to loblolly and shortleaf pines. Recent USFS reports (1972b, 1973) emphasize the continuing destructiveness of the insect. The outbreak area in 1973 included parts of Virginia, North Carolina, South Carolina, Tennessee, Georgia, Alabama, Mississippi, Louisiana, and Texas and two counties in Arkansas. Price and Slentz (1973) estimated that in July and August 1973 there were 1,139,000 infested trees plus 552,000 recently killed trees on a 46-county area of central and north Georgia. Seven of the counties were designated as disaster areas.

Life History

Dixon and Osgood (1961) reviewed the status of the southern pine beetle as a pest and described its life history. The number of generations per year varies principally with climatic conditions; that number is difficult to determine for individual populations because the beetles attack in a complex, overlapping pattern, and may attack almost continuously during favorable weather. Two to three generations per year were reported in the northern portions of the range. The number of generations farther south is greater; Thatcher and Pickard (1967) reported seven generations in Texas.

The beetles overwinter in all stages of development and begin to emerge when climatic conditions are suitable. Emerging adults may attack nearby uninfested trees, or they may migrate a considerable distance. Initial attacks on healthy trees sometimes cause a profuse flow of sap that may "pitch-out" the beetles. If sap flow is only moderate, the beetles move directly through the outer bark to the wood and construct galleries in the inner layers. Eggs are laid along these galleries, and the larvae develop in short larval galleries leading into the phloem region. Fronk (1947) reported, from Virginia, a minimum of 40 and a maximum of 54 days for development from egg to adult. Thatcher and Pickard (1967) reported, from Texas, summer generations of as few as 26 days.

The southern pine beetle has a tremendous potential for increase from generation to generation and season to season. This capacity for increase explains why the outbreaks appear so suddenly when environmental conditions are favorable for the beetles. The precise causes of the rapid changes in population size are not thoroughly understood.

Ecological Effects

These insects characteristically attack trees in massive numbers, so that an infested tree almost always dies. Mortality occurs rapidly, because of the associated penetration of blue-stain, *Ceratocystis minor*. The tree dries quickly, phloem is destroyed, and the foliage begins to turn yellow in about 3 weeks.

Gara and Coster (1968) reported on the aggregation by the beetles, which is brought about by a beetle-produced attractant (Vité et al. 1964). The intensification of infestations is dependent upon synchrony between brood emergence and production of the attractant by other beetles in freshly attacked trees. Their study showed that the distance from which beetles are attracted is limited, but the material is highly effective in aggregating populations in massive attacks on trees.

Many factors related to host conditions are known to predispose trees and stands to attack. Drought and rainfall deficiencies have long been recognized as important factors (Dixon and Osgood 1961) and, in general, there is no doubt that stress brought on by drought may be associated with beetle attack. Also documented is the collapse of outbreaks during periods of excessive rainfall. There are reported exceptions to both of the above generalizations.

Baker (1972) states that outbreaks appear to be caused by conditions that favor an increase in vigor and size of beetle populations while reducing vigor of host trees. Drought, overstocked stands, stand disturbances, and a reduction in abundance and effectiveness of natural enemies of the beetles may contribute to the development of an outbreak. All of these factors may be interrelated. Moore and Thatcher (1973) presented a thorough review of these factors as they relate to population fluctuations of the insect.

When outbreaks are developing, scattered small groups of trees in dense stands are generally attacked. Trees under 15 years of age and less than 2 inches in diameter are rarely attacked at this stage. Once the outbreak is fully developed, however, all age classes and stand densities may be vulnerable. Spot infestations vary in size from single infested trees to an outbreak area of several hundred acres.

Although several predators and parasites have been shown to affect population numbers, it appears that climatic factors are more important in regulating populations. Low temperature can kill a large proportion of overwintering broods (Beal 1933). The duration of outbreaks appears to depend, at least in part, on climatic conditions and their effects on the vigor of the pests as well as of the trees.

Control

Kucera and Barry (1973) reported that the loss of sawtimber in 1973 would exceed 580 million board feet, valued at about $60 million; the infested acreage from 10 states included some 47 million acres of pines. These authors note several suggestions for dealing with the problem. Prompt harvest of infested trees is still the most practical method of treating the problem. Most often this procedure is merely a salvage operation, although with enough manpower to harvest all currently infested material some control could be achieved.

Chemicals are seldom used except for specific purposes where values are high--for example, to protect high-value trees around homesites and recreation areas or to protect seed orchards threatened by attack. Toxic oil sprays that penetrate and kill beetles in the bark are the most commonly used method of application; lindane is the recommended pesticide.

Management practices implemented to maintain and improve the vigor of pine stands are the best procedures for regulating the pest populations, according to current recommendations (Moore and Thatcher 1973). Such practices include the removal of high-risk trees, thinning of overdense stands, harvesting of overmature stands, selection of well-adapted species for plantations, and avoidance of planting on poor growing sites. There is a need for a much expanded research effort on all of the ecological aspects of the problem. Bennett (1971) discussed some of the ideas related to the silvicultural control of the pest. Lorio and Hodges (1968) presented data on moisture stress in southern pines that directly relate to the ecological relationships described above.

Above all, it should be realized that this native insect will cause outbreaks from time to time as long as southern pines are growing. The land manager must learn to cope with the problem and to operate with the beetle as a recognized part of the southern pine ecosystem. In natural succession, pure pine stands are gradually changed to mixed forests and possibly to hardwoods by various processes. The southern pine beetle is one of the ecological factors involved, and occasional outbreaks must be expected in southern pine forests.

Salvage operations to use the trees killed by the current outbreak have been increased. However, the available labor supply and processing facilities have not been able to handle the volume of wood produced. If the outbreak continues, there could be serious wood shortages at some future time because of losses now of young pine forests that would have reached maturity in another 10 to 15 years. Much research is needed on the causes of outbreaks and the procedures necessary to minimize the devastation resulting from beetle infestations.

In 1974 the General Assembly of Georgia passed a new law called the Forest Insect and Disease Control Law (Anonymous 1974). This progressive step is one of several needed to assure that the pine forests of Georgia will be managed for continuous production. The law allows the Georgia Forestry Commission to work directly with landowners in efforts to control the beetle populations. The legislation mandates an ongoing program of detection, protection, and education.

The widespread depredations by this pest of a resource of major importance have been of concern to landowners throughout the South. A very effective advisory group has been formed, and through its efforts and the efforts of USDA administrators and research personnel, a long-term expanded intensive research effort is being initiated. This effort is one of the three expanded special forest insect projects recently approved. This project is planned for a 5-year period and will cost around $20,000,000. By the end of this period a reasonably effective set of forest and pest management systems should be available to land managers.

REFERENCES

Anonymous (1973) Dean Sherman's Forest Industry Affairs Letter. Portland, Oregon, October 31.

Anonymous (1974) Forest insect and disease control law passed. Ga. For. 29:3.

Baker, W. L. (1972) Eastern Forest Insects. U.S. Forest Service, U.S. Dept. Agr. Misc. Publ. 1175. Washington, D.C.

Beal, J. A. (1933) Temperature extremes as a factor in the ecology of the southern pine beetle. J. For. 31:328-336.

Benedict, W. V. (1950) Changes in Blister Rust Control Practices in the Pacific Coast Region. Unpublished Report. Berkeley, Calif.: U.S. Forest Service.

Bennett, W. H. (1971) Silvicultural techniques will help control bark beetles. Soc. Am. For. Soc. Reg. Tech. Conf. 1971:289-295.

California Forest Pest Control Action Council (1969) Forest Pest Conditions in California. Westport.

Campbell, R. W. (1967) The Analysis of Numerical Change in Gypsy Moth Populations. Forest Sci. Mimeograph 15.

Campbell, R. W. (1973) Forecasting gypsy moth egg-mass density. U.S. Forest Service, U.S. Dept. Agr., Research Paper No. NE-268.

Ciesla, W. M. (1973) (Coordinator, Environmental Services, Division of State and Private Forestry, U.S. Forest Service, Missoula, Montana). The Douglas-Fir Tussock Moth: Status, Impact, and Plans for an Action Program in the Western United States. 3rd Draft, October 29.

Clement, R. C., and J. C. T. Nisbet (1972) The Suburban Woodland: Trees and Insects in the Human Environment.

Conservation Rep. No. 2. New York: National Audubon Society.
Collins, S. (1961) Benefits to understory from canopy defoliation by gypsy moth larvae. Ecology 42:836-838.
Detwiller, S. B. (1921) Status of white pine blister rust control in 1920-21. U.S. Dept. Agr. Plant Dis. Bull. June 15, suppl. 17.
Dixon, J., and E. A. Osgood (1961) Southern Pine Beetle, A Review of Present Knowledge. Southeastern For. Exp. Sta. Paper No. 128.
Doane, C. C., and P. W. Schaefer (1971) Aerial Application of Insecticides for Control of the Gypsy Moth: With Studies of Effects on Nontarget Insects and Birds. Conn. Agr. Sta. Bull. No. 724.
Dunbar, D. M., H. K. Kaya, C. C. Doane, J. F. Anderson, and R. M. Weseloh (1973) Aerial Application of *Bacillus thuringiensis* against Larvae of the Elm Spanworm and Gypsy Moth and Effects on Parasitoids of the Gypsy Moth. Conn. Agr. Exp. Sta. Bull. No. 735.
Federal Register (1974) Vol. 39(44).
Fredsall, R. M. (Western Wood Products Association) (1973) Remarks to Panel on Implications of the Recent Tussock Moth Experience to Future Control Operations, Western Forest Pest Committee of Western Forestry and Conservation Association, San Jose, Calif., Dec. 5.
Fronk, W. D. (1947) The Southern Pine Beetle--Its Life History. Va. Agr. Exp. Sta. Tech. Bull. 108.
Gara, R. I., and J. E. Coster (1968) Studies on the attack behavior of the southern pine beetle. III. Sequence of tree infestations within stands. Contrib. Boyce Thompson Inst. 24(4):77-86.
Graham, D. A., and K. H. Wright (1973) Forest insect pest management. Paper presented at the Annual Meeting of the Northwest Forest Pest Action Council, Portland, Oregon, Nov. 8.
Hopkins, A. D. (1899) Report on Investigations to Determine the Cause of Unhealthy Conditions of the Spruce and Pine from 1880-1893. W. Va. Agr. Exp. Sta. Bull. 56.
Kegg, J. D. (1971) The Effects of Repeated Gypsy Moth Defoliation on Oak in the Newark Watershed of New Jersey. Trenton: NJDA Division of Plant Industry.
Kucera, D. R., and P. J. Barry (1973) Southern pine beetle at epidemic proportions. For. Farm. 33(2):16-17,34.
Larsen, A. T. (1973) Oregon Dept. of Forestry. Address to Panel on Implications of the Recent Tussock Moth Experience to Future Control Operations. Western

Forest Pest Committee of Western Forestry and Conservation Association, San Jose, Calif., Dec. 5.

Lewis, F. B., and D. P. Connola (1966) Field and Laboratory Investigation of *B. thuringiensis* as a Control Agent for Gypsy Moth, *Porthetria dispar* (L.). U.S. Forest Service Res. Paper NE-50.

Lorio, P. L., and J. D. Hodges (1968) Oleoresin exrudation pressure and relative water content of inner bark as indicators of moisture stress in loblolly pines. For. Sci. 14:392-398.

McManus, M. L. (1973) The role of behavior in the dispersal of newly hatched gypsy moth larvae. U.S. Forest Service Res. Paper NE-267.

Martin, J. E. (1944) Ribes eradication effectively controls white pine blister rust. J. For. 42:255-260.

Moore, G. E., and R. C. Thatcher (1973) Epidemic and endemic populations of the southern pine beetle. U.S. Forest Service Research Paper SE-16.

Payne, B. R., W. B. White, R. E. McCay, and R. R. McNichols (1973) Economic Analysis of the Gypsy Moth Problem in the Northeast. 11. Applied to Residential Property. U.S. Forest Service Research Paper NE-285.

President's Advisory Panel on Timber and the Environment (PAPTE) (1973) Report. Washington, D.C.: U.S. Government Printing Office.

Price, T. S., and H. Slentz (1973) Evaluation of Southern Pine Beetle Infestations in Georgia. Macon: Georgia Forest Commission Annual Aerial Surveys.

Reese, C. D., and D. L. Becker (1972) The Movement and Impact of Pesticides Used in Forest Management on the Aquatic Environment and Ecosystem. Pesticide Study Series No. 7. Washington, D.C.: Environmental Protection Agency.

Snell, W. H. (1928) Blister rust in the Adirondacks. J. For. 26:472-489.

Thatcher, R. C., and L. S. Pickard (1967) Seasonal development of the southern pine beetle in East Texas. J. Econ. Entomol. 60:656-658.

Tierney, G. C. (1947) Death of Trees Following Defoliation by Gypsy Moths in Connecticut Valley Town of Massachusetts. U.S. Dept. Agr., Bur. Entomol. and Plants Quart. Progress Rep.

Turner, N. (1963) The gypsy moth problem. Conn. Agr. Exp. Sta. Bull. 655.

U.S. Environmental Protection Agency (1973) Environmental Facts, The Tussock Moth. Washington, D.C.: U.S. Government Printing Office.

U.S. Forest Service (1972a) Forest Statistics for the U.S. by State and Region, 1970. Washington, D.C.

U.S. Forest Service (1972b) Forest Insect and Disease Conditions in the United States, 1971. Washington, D.C.

U.S. Forest Service (1972c) The Outlook for Timber in the United States, A Preliminary Summary of the 1970 Timber Review. Washington, D.C.

U.S. Forest Service (1972d) Fact Sheet: Douglas-Fir Tussock Moth Situation. Pacific Northwest Region, Washington-Oregon, Oct. (revised Nov.). Mimeo.

U.S. Forest Service (1973) Forest Insect and Disease Conditions in the United States, 1972. Washington, D.C.

Vaux, H. J. (1954) Economics of the Young-Growth Sugar Pine Resource. Calif. Agr. Exp. Sta. Bull. 738.

Vite, J. P., R. I. Gara and H. D. von Scheller (1964) Field Observations on the Response to Attractants of Bark Beetles Infesting Southern Pines. Boyce Thompson Inst. Contrib. 22:461-470.

Waters, W. E. (1971) Ecological Management of Forest Insect Populations. Proceedings of the Tall Timbers Conference on Ecological Animal Control by Habitat Management. Vol. 3, pp. 141-153.

Waters, W. E. (1973a) Forest Pest Management, A Research Challenge. Agr. Sci. Rev. 11(2):24-28.

Waters, W. E. (1973b) Legal, social and economic constraints on past, present and future pest management practices, pp. 9-22. Pest Management in the 21st Century. Moscow: Idaho Research Foundation.

Wickman, B. E. et al. (1973) Major Outbreaks of the Douglas-fir Tussock Moth in Oregon and California. USDA Forest Service Tech. Rep. Pacific Northwest Forest Range Experiment Station. PNW5.

5

TRENDS IN FOREST PEST MANAGEMENT: SOME PROJECTIONS AND SUGGESTIONS

Over the past 30 years, notable changes have taken place both in the nature of forest management in the United States and in the extent and kind of forest pest control practices employed in such management. Conscious efforts to increase wood growth by intensified cultural methods have expanded significantly. Recreational use of forests has virtually exploded. Preservation of certain forest areas in their primitive condition has become an explicit and important goal in public forest policy. Technology for forest pest control, whether by chemical or biological means, has begun to emerge.

It is certain that equally significant changes affecting forest pest control will take place during the next 30 years. Any attempt to predict such future changes in specific and quantitative terms would, given the present state of our knowledge, be little better than speculation. Nevertheless, it is possible to identify certain trends in forest management and in pest control that seem almost certain to make the future different from the past. If such trends provide an inadequate basis for writing the scenario for the next 30 years, they still furnish some clues to the way in which the plot is likely to unfold.

We assume that over the next three decades, population in the United States will increase to perhaps 35 percent above the present level. There are finite limits to the land base, and the larger population will require more land for nonforest purposes; thus, the area in forests will almost surely decline somewhat, probably by about 1 percent per decade. In other words, a significantly larger population will have to derive the total array of forest-based goods and services that it will need from a somewhat smaller area than that available today.

For the environmental and service functions of forestry, this diminution in size of resource base appears to be absolute. In the case of wood products, however, there is the possibility that expanding future demand could be met by increasing net imports of lumber, plywood, pulp, or other wood materials. The world wood supply has been carefully studied by a variety of agencies in recent years (USFS 1973, Zivnuska 1967, UNFAO 1966, PAPTE 1973), and similar conclusions have been reached. The growing pressure of expanding wood demand against relatively limited supply is as characteristic of the world wood market as it is of the United States. There is little realistic prospect of this country's being able to rely on foreign forests to meet a larger proportion of its wood needs. On the contrary, with respect to the coniferous wood, which accounts for three-fourths of current U.S. consumption, there are strong reasons to argue that this country's policy should be to decrease its relative and absolute reliance on foreign sources by increasing its own softwood exports. Such reasoning recognizes that the United States ranks third among the countries of the world in area of coniferous forest; that it has more favorable conditions for rapid conifer growth than any other country; and that, in the light of world wood markets, expanded exports of U.S. coniferous wood would, if domestic supply potentials were fully utilized, contribute significantly in the long run to the solution of this country's balance of payments problem.

Thus, merely to maintain present levels of living, forest land will have to be used more effectively (whether for wood, wilderness, water supply, or wildlife) than it is today. If standards of living are to continue to rise in the future as they have in the past, the pressure to use the forest more effectively will be even greater.

MORE INTENSIVE MANAGEMENT

"More effective use," as employed here, means getting a larger yield per acre of whatever consumer benefits may be involved. In the case of wood product benefits, it might mean producing more harvestable wood per acre per unit of time; and in the case of wilderness values, it might mean providing for more man-days of wilderness recreation per acre per unit of time without loss of quality of experience. In general, assuming that the forest is to some degree heterogeneous, more effective use will depend on basing the management of the land on fuller information and better understanding about the forest itself and about its use. The amount of information that can be introduced into the management process is an inverse function of the minimum-sized unit of land chosen as the basis for management planning. For example, a management plan for a 10,000-acre tract based on one thousand 10-acre units will require far more information than a management plan based on ten 1,000-acre units.

The costs of managing the land will be higher, of course, the smaller the units. As the value of the resource increases, however, it becomes profitable to manage it in smaller and smaller units: the benefits of planning in these smaller units on forest land are expected to rise faster than the costs. The consumer benefits mentioned above, then, will actually be net benefits.

Thus, over the long term, the necessity for more effective use of the land will require that management planning move steadily in the direction of substituting planning by the acre for planning by the section or by the township. This tendency is fundamental, whether we are discussing timber production, the provision of wilderness recreation, or some other forest use. The additional understanding needed for more effective use is, of course, essentially a function of research. Thus, although there is some evidence that we are not now applying in the management process all of the research information currently available, any "surplus" research that now exists is likely to be rapidly absorbed as management moves to a smaller and smaller unit as a basis for planning. Indeed, lack of sufficient research information is likely to be a limiting factor on how rapidly planning can, in practice, be based on significantly smaller unit areas.

A second potential limitation on the needed intensiveness of management is the availability of a sufficient number of resource managers with an adequate level of professional preparation and competence. Although reliable data are unavailable, the ratio of professional manpower to area of forest land in the United States is believed to be only one-fourth to one-third of that in countries like Germany or Switzerland, where forests are intensively managed. The need for more thorough professional training has been pointedly noted by forestry educators (Dana and Johnson 1963). It is also implied by the gradual extension of state legislation requiring professional licensing for foresters.

A second trend, correlated with but still distinct from the first one, is the almost certainly increasing real value per unit of the various outputs from forests.

Increasing scarcity of timber, wilderness, recreation, wildlife, and other classes of forest value has characterized the past three decades. The index of such increased scarcity has been the increase in relative exchange values of such goods and services. Established population trends and the relative stability of tastes lend high probability to continuing significant value increases. This means that greater resource management costs (i.e., managerial inputs) will be justified (and probably forthcoming) to increase average annual yields per acre. As the permissible level of cost per acre for management rises, more complicated management prescriptions can be accommodated.

In addition to these trends is the fact that the specific costs for most pest control measures tend to vary with the number of acres treated, rather than with the volume of goods or services protected by the treatment. If yields of wood per acre (or of recreation or any other forest benefit) are increased, the pest control cost per unit of yield tends to fall. Thus, as the unit value of yields rises, methods of pest control previously regarded as uneconomic may become profitable.

In sum, the foregoing trends mean that both the incentives and the means for managing forest pests more intensively will increase steadily in the years ahead. Large owners, both public and private, may be expected to respond positively by managing for increased output of valuable goods and services; but on the large proportion of forest area in small private holdings, the outcome is less clear. The time perspectives and land use preferences of these landowners may well preclude intensive management unless new forest policy initiatives are effective in overcoming owner apathy. Wherever rising incentives encourage more intensive management, careful, scientifically meticulous control methods will be more necessary, because greater values will be at stake, more money will probably be available to finance pest control efforts across a much broader array of pest problems, and the opportunity costs of *not* managing pests will be higher. These same expansions in both incentives and means for pest management could also substantially increase the dangers of wider use of new technologies or pesticides without adequate knowledge of their full biological ramifications. Under such circumstances, the need to discriminate more carefully between those situations where a given organism should be regarded as a pest and those where it should not be so regarded will grow.

CONSERVATION OF ENERGY

Compared with other basic structural materials and plastics, wood is a comparatively light user of harnessed energy in the growth and primary extraction phases and (with the notable exceptions of pulp, paper, and cellulose products) in the fabrication and installation phases of its use (Dane 1972). Because most wood products are biologically degradable over time, wood waste disposal seems likely to be less energy-consuming than disposal of mineral wastes. If, as seems probable, energy supplies continue to become more restrictive and society requires more rational disposal of solid wastes, the use of wood materials will offer increasing social advantages over more energy-demanding structural substances.

At the same time, economic pressures are likely to work against the energy-intensive practices of forest management and for less energy-intensive ones. For example, controlled use of fire for planting site preparation might be favored over use of heavy equipment for this purpose. However, little work has been done to assess the comparative efficiency of different practices under tightened energy constraints.

CHANGING TECHNOLOGICAL AND SCIENTIFIC VIEWPOINTS

In addition to the economic factors just discussed, both the long-run effectiveness of forest pest control and the nature of its environmental impacts are strongly affected by the character of the control technology available. Such technology is, in turn, largely an outcome of past research and development efforts in the agricultural sciences as well as in forestry. It seems clear that from about World War I to at least the mid-1960s most forest insect and vertebrate control research and development was aimed at acquiring more market-cost-efficient techniques for eliminating pests. A variety of evidence now demonstrates that such an approach misses the real nature of most forest pest problems and is totally inappropriate as the dominant R&D strategy.

Both the reorientation of much pest research toward more rational goals and the hastening of progress toward acceptable and effective control systems are handicapped by several attributes of the total research enterprise. The first of these is the diffuse nature of the overall forest pest problem. The disciplines involved--entomology, zoology, plant pathology, and weed science--are in large measure separate from forestry, and, as already mentioned, forests provide no single commodity as an agreed-upon common focus and measure of research achievement.

Second, the administrative structures through which pest control research is carried out often serve as a deterrent to major shifts in research objectives. The pesticide industry obviously had a commitment to particular approaches and provision of materials. Its R&D effort is now in some disarray and almost certainly will withdraw from many markets that are too controversial or too small to warrant the extremely high cost of development and registration. The regional forest experiment stations of the USFS and the state experimental stations and forestry schools supported with both state and federal funds have a vast variety of commitments, ranging from straightforward chemical control methods to population biology and epidemiology of pest species. But interest and missions of the other agencies are also involved--notably the forest industries, the Bureau of Sport Fisheries and Wildlife of the U.S. Department of the Interior, and the National Science Foundation through its support of basic ecology and ecosystem research, as well as individual investigators supported by foundations and industry. All these are at times concerned with the biology of host and pest species, often specifically with their interactions, yet there has been only negligible effort at coordination among these agencies, except through the initiative of the individual investigators.

Third, as a result of the foregoing, the R&D efforts applicable to pest problems are diffused. Unless a crisis develops, there has been little incentive for a single agency to concentrate or cooperate intensively with others on a single pest problem.

Finally, research funding is entirely inadequate in view of the overall magnitude of the pest problem and the high costs involved in developing new materials or new approaches even for the major forest pests. The reduced participation of the pesticide industry in research and development has not been compensated by any proportionate increase in the funds available for alternative approaches. Yet the development of alternatives will require extensive research efforts, including, for example, thorough study of pest population biology, life table studies, isolation and synthesis of attractant substances, and the comprehensive investigations required to formulate systems of "integrated control," all of which are time-consuming. In addition, they cost money, which until now has not been provided in adequate amounts or in timely fashion. For example, a USFS research director commented: "We spent $370,000 on tussock moth research last year, and we needed four or five times that amount. . . . The irony is that we made more progress on this pest than on any other major forest pest in the country. We did a lot with a very little money. We are on the verge of finding alternatives to DDT."

DEVELOPMENT OF SYSTEMS OF INTEGRATED PEST MANAGEMENT

The diffuse nature of the forest pest control effort as described above has not gone wholly unnoticed by the USFS, its regional offices, or other government agencies. As in agriculture, an increasing need has been felt in forestry to deal in a more systematic way with the losses caused by insect pests and diseases (so far, the effects of competing vegetation or vertebrate pests have not been explicitly included in most of the new approaches). Fragmentation has begun to be recognized as a serious problem in research, in evaluation, and in the development of control programs.

A potential solution to these problems may be through the approach called "integrated pest management." Although this term is something of a catchall, with no precise definition, it has been described by the Council on Environmental Quality (CEQ) as "an integrated and comprehensive approach to the use of various control methods that takes into account the role of all kinds of pests in their environment, possible interrelationships among the pests, and other factors" (CEQ 1972).

The *concept* of integrated pest management in forestry is not really new. For the reasons described earlier in this report--ecological diversity and complexity, wide-ranging social and economic values, and differing management objectives--forest managers have long recognized that a variety of methods and a highly flexible planning base are needed to cope effectively with destructive pests on a continuing, long-term basis. Conceptually, also, they have recognized that the actions taken should be fully integrated with other management activities to achieve specific objectives. In practice, however, this concept has never been fully implemented. This is partly the result of the attitudes of forest resource managers and pest control specialists, most of whom have been concerned primarily with insects that occur in outbreaks and with other pests that cause obvious damage. The tendency has been to wait until a "serious" problem develops, and then to make short-term decisions on what to do about that specific and immediate problem.

Another constraint on the development of pest management systems has been a lack of knowledge and technology for assessing the full impacts of pests on resource productivity and values. This area, impact assessment, represents perhaps the weakest point in present forest pest control activities. It is vital to develop a better understanding of the difference between ecological effects and socioeconomic impacts of destructive pests and the interaction of the pests' impact with management objectives. This problem area was discussed in depth at a meeting of USFS and state pest control personnel at Marana, Arizona, February 7-11, 1972. The recommendations for action developed at this workshop emphasized the need to expand pest survey coverage and to improve data collection and analysis techniques (USFS 1972).

Also lacking are the technologies and information for predicting the occurrence and/or buildup in numbers of and damage from insect or disease pests and for evaluating benefits and costs of alternative approaches to managing them. For reasons cited earlier, including increased pressure for greater productivity and usefulness of forest lands and environmental concerns, it is becoming apparent that a more comprehensive, interdisciplinary, and larger-scale research effort will be needed to solve these problems.

The problems have been recognized, at least by some, and a start has been made in developing potential solutions. Several of the ongoing or developing forest pest management research programs have been mentioned in our case studies (see Chapter 4). These include the USDA gypsy moth R&D program; the expanded USFS-CSRS (Cooperative State Research Service) R&D program on the southern pine beetle; and a similar program on the Douglas fir tussock moth. Significant changes are also being made, on a smaller scale, in USFS research programs on dwarf mistletoes and fusiform rust of southern pines. Several other programs are under development.

One of the largest-scale research efforts in this field at present is the International Biological Program project entitled "Principles, Strategies, and Tactics of Pest Population Regulation and Control in Major Crop Ecosystems" (described briefly in Volume I of this Study), which is funded jointly by the National Science Foundation and the Environmental Protection Agency. The pine bark beetle subproject, which includes study of three major complexes of bark beetles and their hosts, has been a focal point in coordination of ongoing research on this topic, as well as providing some additional funds for research on the development of pest management approaches to control of these species.

All of the above-mentioned programs place primary emphasis on research. Making integrated pest management a reality in terms of actual operations and control is likely to be a long and difficult process.

Region 5 (California) of the USFS has been quite active in attempting to correct the problems created by the fragmented nature of forest pest control, research, and evaluation. The USFS in this region is developing a coordinated forest pest evaluation system, which aims "to place pest control clearly within the context of resource management, and to bring to the task of evaluating insect and disease depredations a coordinated, systems approach" (MacGregor et al. 1973). It is proposed that pest control be approached through the concept of "pest complexes." An example of a situation where the need for such an approach is evident is the bark beetle problem: with many species of bark beetles, numerous biotic and abiotic stresses may interact to predispose a tree to attack by these major pests.

It appears that the research and operational advances outlined above represent a trend toward the development of more rational integrated pest management systems. These are still far from accepted or commonplace, however, and efforts will be necessary to ensure that these and similar programs continue to receive support.

INTRODUCTION OF EXOTIC SPECIES AND ATYPICAL EXPANSION OF NATIVE PESTS

History has clearly shown the importance for pest problems of the introduction of exotic trees and pests and of atypical behavior of native pests. Similar developments will undoubtedly occur in the future. Such developments may be quite significant in the areas discussed below.

Exotic Pests

The introduction of foreign pathogens and insects sometimes results in great damage to native plants. The previous separation of host and pest precludes evolutionary development of resistance or adaptation by the plant; also, the new organism often arrives without the parasites and predators that affect it in its homeland. Either of these factors may permit rapid or efficient expansion of the pest population. Our most dramatic and devastating forest tree diseases, including chestnut blight, Dutch elm disease, and white pine blister rust, are all caused by introduced pathogens. Introduced insect pests include the gypsy and brown tail moths, European pine shoot borers, and sawflies. The efficiency and quantity of international travel and commerce suggest that the threat of exotic pest introduction will remain high. The problem of introduced weeds is usually of less importance in forests, although Kudzu and Japanese honeysuckle--both purposeful introductions that have escaped--are aggressive lianas that seriously interfere with tree growth in parts of the Southeast.

Exotic Plants

If a tree is planted outside its natural range, it becomes an exotic and may be vulnerable to insects and pathogens for which it has no inherent resistance. Attempts to introduce pines to the Great Plains for windbreaks, for example, have encountered numerous pathological problems. The use of forest species in urban areas has frequently been very costly and only partially successful because of pathological stress and insect pest problems. In spite of the long-standing awareness (Boyce 1954) of the risk of planting "offsite," species are still removed from their natural habitats and introduced into unnatural situations.

Intensive Cultural Practices

The technique of planting large areas with trees of a single species and of similar age and pruning, thinning, fertilizing, or irrigating them has occasionally led to increased insect and disease damage. Diseases that have been enhanced by cultural practices imposed by man include fusiform rust of loblolly and slash pine (Miller 1972), annosus root rot of red pine (Tegethoff 1973), and stem cankers of American sycamore (Ross 1971). As it is reasonable to anticipate increasing acreages of plantation forests and an increasing level of "cultural" attention to plantations and other forests, it is rational to anticipate increased insect and disease problems associated with these activities (Shea 1971).

Expansion of Native Pests

Occasionally a native pathogen or insect will increase its population and cause a significant problem on a native tree species. Subtle changes in climate or surrounding plant and animal populations may be involved. The increased significance of comandra blister rust (caused by *Cronartium comandrae*) on lodgepole pine in the first half of this century (Krebill 1965) and its recent expansion on loblolly pine (Cordell and Knighten 1969, Wolfe 1970) are examples of this incompletely understood process. Oak wilt and birch dieback are other examples of this phenomenon. It is difficult to predict the future significance of a phenomenon as poorly understood as atypical expansion of native pests.

Conclusion

We may conclude that the possibility of occurrence of particularly damaging pests remains approximately the same as it has been in the recent past in regard to the introduction of exotic pests and atypical expansion of native pests. In regard to pests associated with intensified cultural activities and planting of exotic trees, an increase in significance is probable.

DEVELOPMENT OF NEW PEST CONTROL TECHNOLOGIES

Although the control technology available today is largely an outcome of past research and development efforts in both agriculture and forestry, there are suggestions that new technologies may soon be developed to the point of operational usage.

Classical biological control techniques, in which parasites or predators of nonnative pest species are imported and established, have been fairly successful. Mention is made elsewhere of the use of *Bacillus thuringiensis* against gypsy moth and forest tent caterpillar. In addition, foliar sprays of polyhedrosis viruses have given effective control of some sawflies, indicating possible commercial use in the future. Such techniques have not been well explored in forestry. An expanded research effort should include examination of a broad array of organisms feeding on insect pests, such as birds, shrews, other insects, bacteria, fungi, and viruses.

Little is known about the use of microbial agents against fungal infections, although their importance in natural regulation is great. Rishbeth's (1970) finding that *Fomes annosus* is excluded from freshly cut stumps by simultaneous or prior inoculation with *Peniophora gigantea* has led to several field trials of the latter fungus as a protective or pre-emptive agent. Thus far, the practical consequences seem equivocal, in part because parallel treatments of stumps with simple chemicals are at least equally effective. Neither treatment has yet gained real operational significance because of the cost.

To combat native pests, there is considerable interest in encouraging the natural enemies that are already present, by providing refuges or alternative food sources. The practices of intensive culture tend to suppress plant diversity, but they may be modified to favor particular understory and edge species if there were demonstrable evidence of their contribution to pest regulation. For example, intensive site preparation measures are known to increase, temporarily, the abundance of native legumes in the Southeast, and at least some of these could be maintained by recurrent fire, or by application of nonnitrogenous fertilizers. Again, the adults of some parasitic wasps require particular nectar sources that might be maintained either in the understory or in the network of minor openings and road edges characteristic of intensive culture areas.

Purposeful mixture of overstory species poses several problems, not only of site adaptability but especially of management and harvest; this method has not yet been found acceptable. Nevertheless, "block-wise" mixtures or juxtaposition of unlike species in narrow bands or other configurations that still allow ready management and harvest are not at all unthinkable if clear evidence of increased pest resistance were to become available.

Thus far, however, the only current evidence bearing on species mixtures concerns the possible influence of red alder upon the serious root rot of Douglas fir, *Poria weirii*. Development of this fungus is inhibited by substances--possibly phenolic substances--in the roots of the alder. Though not yet shown to be effective in practice, admixture of alder with Douglas fir is suggested as a means of reducing damage from this fungus. Alder is a major weed species, because its vigorous early growth suppresses young Douglas fir and other conifers. Yet some compatible mixtures are known and might be duplicated under intensive culture if the increased management effort were warranted. Furthermore, the high nitrogen-fixing capacity of alder is notably advantageous to companion species on low-nitrogen sites.

Other specialized techniques using biological principles also show some promise, including the use of pheromones in some preliminary control studies. To date, these techniques appear to be most useful as survey tools. This is, of course, an important role, as better evaluation techniques are much needed. Manipulation of bark beetle populations using attractant chemicals have also shown promise, as mentioned in the southern pine beetle case history (see Chapter 4).

INSTITUTIONAL CONSTRAINTS AFFECTING PEST CONTROL

Two classes of future changes in state and federal policies may affect the trend of forest pest control. One would involve changes that affect forest management generally and thus have impacts on pest control as an element in forest management. On private land, the recent trend toward more rigorous state regulation of private timber-harvesting practices in the interest of fuller and faster restocking of the land is a case in point. To the extent that such laws become more widespread and effective, private owners in the future will, on the average, have better-stocked young forests than they have had in the past. With the greater value of the timber on their lands, owners may be expected to press more urgently for the protection of their forests than they have in the past. The same increase in values may justify extension of protection efforts.

There is also a current policy thrust to secure more intensive management of national forests and other public land, either through earmarking of timber sale receipts for this purpose or by some other mode of financing. If such a policy of intensified national forest management is adopted by the Congress, it, too, will almost certainly lead to expanded pest control efforts. The sugar pine case described above demonstrated the illogic of spending large sums solely to protect young forests from disease in the absence of measures to encourage rapid timber growth. It would be equally illogical to spend large sums to increase national forest timber growth without commensurate expenditures to protect such growth, especially from insect, disease, and mammal damage.

The second class of prospective public policy changes would include further measures to internalize or control costs of environmental impacts: e.g., proliferation of bans on the use of pesticides and adoption of more rigorous standards of air and water quality. To the extent that institutional changes of this sort emerge, significant impacts on forest pest control are to be expected; but it is difficult to assess their probable direction. On the one hand, there are some situations where management could respond to such changes by "improving on nature" through wider application of biological controls and silvicultural approaches to develop forests with increased resistance to certain pests. Although this concept, based on the maintenance of a healthy and dynamic forest conforming closely to a natural ecosystem and with wood and other products as readily derivable by-products of that condition, may occupy a larger role in the forestry of the future, it seems at best to be only a partial solution. Pests are defined by what we want from the forests, and as long as those wants continue to expand, the pressures to use such practices as genetic selection, uniform stands, and pesticides will increase. Retreat from sensible exploitation of these techniques seems unlikely and unnecessary. When further development in these directions occurs, then, further regulatory changes to protect environmental quality will tend to place the costs of environmental impacts of pest control on users of forest goods and services, but they will not eliminate those costs.

SOME IMPLICATIONS FOR FOREST PEST PROBLEMS OF THE FUTURE

The preceding discussion of general trends in the forest situation in the United States leads to several conclusions about the changing character of forest pest problems in the decades ahead. Since management objectives are a fundamental determinant of pest problems, we evaluate these trends separately by the primary purpose of management.

Objective: Timber Growing

Most of the forest pest control operations undertaken in the West in the past have been aimed at organisms that were destroying merchantable timber, usually old-growth. There have been positive economic incentives to minimize losses of this kind, which, with a few notable exceptions, were due to pests associated with timber at or close to the stage of decadence. Three factors seem likely to make this sort of pest problem relatively less important in the future than it has been in the past. First, logging has steadily reduced the area of virgin old-growth forests, which are most susceptible to this class of damage. Second, the same process has resulted in establishment of more extensive road systems in or near those old-growth forests that remain; the existence of a widespread network of forest roads may make salvage of threatened or damaged trees more feasible in the future than it has been in the past. Third, those old-growth forests that remain tend to be on poorer or less accessible sites where pressures to reclassify the land from commercial timber to some form of recreational reserve has begun to focus. To the extent that such land is withdrawn from the commercial forest land base, the management objective is in effect changed; and preservation of merchantable timber volume per se no longer ranks as a primary management objective. This could reduce the pressure for pest control on such land.

At the same time, more and more land is being consciously managed more intensively for permanent timber growing, particularly on industrial ownerships. Where young-growth timber management thus becomes the primary goal, the incidence of pests is quite different from that in the old-growth case. Protection of the productive capability of the stand, rather than mere protection of merchantable timber volume, becomes the key management goal. As the management emphasis moves in this direction, several significant changes in pest problems take place. A spectrum of potential pests quite different from those associated with merchantable timber must be considered, as the insects and diseases of juvenile trees are different from those of mature trees. In addition, a number of plants and animals, particularly those that damage tree seed, seedlings, and saplings must be added to insects and diseases as possible pests. The economic pressure to protect young forests from such adversaries becomes much greater, not only because the young trees now have greater value than they did in the past, but also because the trees themselves are not volunteers (as was the case in the young forests of the past) but represent conscious and expensive investments in regeneration, which must be insured against losses.

The future, then, will see fewer forest pest problems characterized by large areas of relatively uniformly aged forests subject to damage from one or a few insect and disease enemies. But it will see a greater variety of problems involving insects, diseases, animals, and weed plants, more locally defined by forest composition, stand age-class, and site conditions. Broadcast treatments are likely to give way to more carefully controlled managerial approaches.

Much wider use of artificial regeneration and genetically selected stock is to be expected. Most forest tree species are characterized by substantial genetic variability, and often by significant differences within species in adaptation to various climatic and site conditions. Major steps have already been taken to insure against increased pest problems resulting from the planting of natural populations in locations to which they are ill adapted, although in some regions there is still room for better use of existing knowledge of seed origins and site adaptations.

The current attention to genetic selection (in the interest of rapid growth, better form and wood properties, or pest resistance) offers the possibility of plantations with reduced genetic variability, and hence, with greater vulnerability to unforeseen pests and stresses in the future. In view of the repeated emphasis of forest geneticists on maintenance of genetic diversity within their selections, however, and the genetic diversity of seed from orchards now coming into production, this possibility may remain more theoretical than real.

Of greater concern may be the concentration upon one or a few species for planting within a region, such as loblolly pine or Douglas fir, and the increase in area of plantations of single species. Artificial regeneration for wood production usually, although not necessarily, entails planting a single species on an area of some size in the interest of efficient management. Quite apart from any biological interactions, the inescapable hazard of extreme biotic or abiotic events, however faint, sets limits to the dependence that a prudent landowner or industry will wish to place on a single species and to the area devoted to its exclusive culture. The long period of growth required to repay the establishment costs and the difficulties of contending with unforeseen biotic events urge greater conservatism than has been applied in agriculture. At present, there is no objective evidence to suggest what limits to such dependence are appropriate, but the effort now given to culture of associated species--western hemlock and Sitka spruce in the Pacific Northwest, and other pines and hardwoods in the South--suggests that the potential limitations of "single crop" intensive forestry are widely recognized.

Objective: High-Use Recreation

Virtually all forms of forest recreation are expected to increase in the future. The explosive growth experienced during the last several decades may be dampened in the future by energy constraints, shifting public tastes, or other factors, but continued growth in forest recreation seems assured.

Certain pest problems in high-use recreation areas are associated with either the extent of such areas or the levels of use or both. An example would be increased susceptibility to insects of trees caused by excessive soil compaction in a campground. Unlike the situation on timber-growing land, there is little reason to expect significant change in the character of this sort of pest problem, although its incidence will rise with increasing use.

A second problem in high-use recreation areas occurs when pests have an adverse aesthetic impact on users but when the pest occurrence is unrelated to the level of use. For example, localized outbreaks of the tent caterpillar, the tussock moth, and other pests may, if they occur in strategic high-use areas, cause anything from minor annoyance to major aesthetic losses. Where annoyance or loss is sufficiently severe, increasing recreational use will mean increasing political pressure for remedial action. In such situations, silvicultural methods of control may be ruled out if extensive manipulation of the vegetation would be required, as this manipulation might itself damage the forest area.

Objective: Low-Use Recreation

The future seems likely to bring least change and least intensification of pest control problems in those forest areas where management focuses on providing recreation with low levels of use (e.g., wilderness areas, most hunting and fishing areas). Under such management, the number of organisms that must be regarded as pests is far smaller than in other cases. Indeed, because a pest is defined by reference to human goals, the definition of pure wilderness can be considered to rule out the existence of any pest organisms. In practice, certain recreation values are almost certain to be held in high regard in wilderness areas, and these may be exposed to pest damage. It appears unlikely that control programs can be justified in wilderness areas, except in the limited number of cases where damage is extensive or where the mobility of the pest is so great that there is significant danger of its doing substantial damage outside the wilderness.

In certain types of low-use recreation, such as hunting and fishing, the values derive mainly from the faunal populations. Disease epidemics among fish and game populations may require control action; but in most cases, these will have little unfavorable impact on the plant cover and, hence, do not raise many of the problems of pest control inherent in other management situations.

Low use is usually accompanied by a low level of surveillance of the resource and, therefore, by limited pest detection capability. In such areas, pests may be detected only after it is too late to initiate fully effective control measures. Low-use recreation ordinarily implies that the forest area will be maintained with a relatively low level of road development (in wilderness areas, by definition, there will be none). For this reason, whenever pest control measures do become necessary, it is likely that they will have to be applied by very extensive means, such as aerial spraying. To the extent that this implies less localized control over the effects of the treatment, combating pest problems may continue to cause spillover and other side effects that have been associated with comparable operations in the past.

Objective: Nursery Production of Planting Stock

The trend toward heavy dependence on planting in production forestry is already well established. In those states where reforestation after logging is required by law, nurseries may furnish the only means by which such requirements can be met in practice. This role for nurseries is likely to become more widespread, but a major expansion in nursery area is not likely.

Because nurseries support enormously valuable crops ($10,000 to $30,000, 1974 basis, per acre at harvest) and because each acre of nursery may furnish seedlings for the reforestation of 500 to 2,000 acres of resource lands, the risk of even small production losses has major implications, which may become more serious in the future.

Managers of the highly intensified nursery cultures have relied principally on fumigation plus a few marginally effective herbicides for pest control, together with much hand labor. Such labor is increasingly difficult to obtain, and the prospects of further commercial development of pesticides for this specific use appear to be small. Therefore, it is likely that nurseries will have increasing difficulty in meeting their commitments, and hence, reforestation programs will suffer. This problem is already forcing industrial owners into reliance on container seedlings, for which there is an entirely new array of pest problems in both nursery and field. The future will probably see the development of a number of new approaches to seedling production, born of crisis, only some of which will be useful. There are likely to be many failures in production, which means that many thousands of acres to which the seedlings were committed will join the reforestation backlog. The inevitable result will be an increase in demand on nursery facilities that already have production problems.

Forest nurseries fall somewhere in an administrative no-man's land between forestry and horticultural research programs. Nursery production receives little attention from either in the area of pest management. Continuation of this status will contribute to the perpetuation of the production difficulties already being encountered.

Objective: Amenities from Urban and Suburban Trees

The value of urban and suburban trees will undoubtedly remain high or even increase in the future. Interest in trees may increase because of their aesthetic appeal and because of our increased appreciation of their amenity functions. Although more and more species are being selected and planted because of their special tolerance of urban-suburban growth conditions, many species are and will continue to be poorly adapted to these unnatural habitats. Pathological and entomological stresses, therefore, will continue to be significant.

Urban tree pest management will probably remain relatively constant. Programs will be determined primarily by municipal and county budgets, past practice, and the competence of existing personnel. New urban insect strategies--for example, the Berkeley, California, "ecosystem insect program" (Olkowski 1974)--will remain the exception. This plan may provide a model for an urban insect control program that employs less pesticide, saves considerable money, and results in less insect damage.

In suburbia, new federal and state pesticide laws will restrict the materials available to the homeowner. Certification and licensing provisions of these new laws will probably force marginal pest control firms out of business. Larger and more diversified commercial operators will be favored. The net effect on the homeowner will be higher pest control costs and, one hopes, more sophisticated advice and more efficient pest control.

REFERENCES

Anon. 1974. EPA permits DDT emergency use against tussock moth. J. For. 72:233-234.

Boyce, J. S. (1954) Forest Plantation Protection against Diseases and Insect Pests. Forestry Development Paper No. 3. Rome: FAO.

Cordell, C. E., and J. L. Knighten (1969) Comandra blister rust on young loblolly pine in eastern Tennessee. J. For. 67:332-33.

Council on Environmental Quality (1972) Integrated Pest Management. Washington, D.C.

Dana, S. R., and E. W. Johnson (1963) Forestry Education in America Today and Tomorrow. Washington, D.C.: Society of American Foresters.

Dane, C. W. (1972) The hidden environmental costs of alternative materials available for construction. J. For. 70(12):734-736.

Krebill, R. G. (1965) Comandra rust outbreaks in lodgepole pine. J. For. 63:519-522.

MacGregor, N. J., J. W. Byler, and D. R. Hart (1973) Forest Pest Evaluation System, an Overview. Preliminary draft. U.S. Forest Service, California Region, Timber Management, Branch of Pest Control.

Miller, T. (1972) Fusiform rust in planted slash pine; Influence of site preparation and spacing. For. Sci. 18:70-75.

Olkowski, W. (1974) Managing an ecosystem insect program. Grounds Maint. 9:41-43, 69-75.

President's Advisory Panel on Timber and the Environment (PAPTE) (1973) Report. Washington, D.C.: U.S. Government Printing Office.

Rishbeth, J. (1970) Possibility of stump inoculation for conifers other than pines [*Peniophora gigantia, Fomes annosus*]. International Conf. *Fomes annosus* Proceedings. 3D:110-120

Ross, E. W. (1971) *Diplodia theobromae* and *Ceratocystis fimbriata* f. *plantani* found in silage sycamore plantings. Plant Dis. Rep. 55:741-743.

Shea, K. R. (1971) Disease and insect activity in relation to intensive culture of forests, pp. 109-118. *In* Proc.

XV IUFRO Congress, March 14-20, 1971. Gainesville: University of Florida.

Tegethoff, A. C. (1973) Known distribution of *Fomes annosus* in the Intermountain region. Plant Dis. Rep. 57:407-410.

United Nations Food and Agriculture Organization (UNFAO) (1966) Wood: World trends and prospects. Unasylva 20(182):1-12.

U.S. Forest Service (1972) Report of the Workshop on Impacts of Insects and Diseases on Uses, Values, and Productivity of Forest Resources. Washington, D.C.: U.S. Department of Agriculture.

U.S. Forest Service (1973) The Outlook for Timber in the United States. Forest Resource Report 20. Washington, D.C.

Wolfe, R. D. (1970) Status of comandra blister rust in Arkansas. Plant Dis. Rep. 54:471-474.

Zivnuska, J. A. (1967) U.S. Timber Resources in a World Economy. Washington, D.C.: Resources for the Future, Inc.

APPENDIX A

REPRESENTATIVE DISEASES IN FOREST AREAS OF THE UNITED STATES

TABLE A-1 Forest Diseases in the North (Lake States, Central, Middle Atlantic, New England)

Host	Disease	Agent	Location	Control
<u>Forest</u>				
Hardwoods				
Oak	Oak-wilt	*Ceratocystis fagacearum*	West Virginia, Pennsylvania	Nothing, cultural
Several (sycamore, oak, ash)	Anthracnose	Several fungi	General	Nothing
Beech	Beech bark disease	*Cryptococcus fagi*, *Nectria coccinea*	Vermont, Massachusetts, New York	Cultural
Aspen	Hypoxylon center	*Hypoxylon prinatum*	Lake states	Cultural
Conifers				
Several	Annosus root rot	*Fomes annosus*	Rhode Island, Connecticut, New York	Cultural
White pine	Root decline	*Verticicladiella procera*	Ohio, New England	Nothing, cultural

Several	Needle disease	Several fungi	Nurseries	Cultural, fungicide
White pine	Blister rust	*Cronartium ribicola*	New York, New England	Cultural
Red pine	Root rot	*Cylindrocladium scoparium*	Nurseries, Michigan, Minnesota	Cultural
Scotch pine, red pine	Needle cast	*Lophodermium pinastri*	Nurseries, Lake states	Cultural, fungicide
Red pine	Shoot blight	?	Minnesota, Wisconsin	--
<u>Christmas trees</u>				
Several	Scleroderris canker	*Scleroderris lagerbergi*	Vermont	Cultural
Several	Needle disease	Several fungi	General	Cultural, fungicide
Black spruce	Needle rust	*Chrysomyxa ledicolo*	Minnesota	Cultural, fungicide
Scotch pine	Needle rust	*Lophodermium pinastri*	Lake states	Cultural, fungicide
Spruce	Needle rust	*Rhizosphaera kalkhoffix*	Wisconsin, Minnesota	Cultural, fungicide
Scotch pine	Needle disease	*Scirrhia acicola*	Wisconsin	Cultural, fungicide

TABLE A-1 (continued)

Host	Disease	Agent	Location	Control
<u>Urban</u>				
Hardwoods				
Elm	Dutch elm disease	*Ceratocystis ulmi*	General	Cultural, pesticides
Elm	Phloem necrosis	Mycoplasm	General	Cultural
Elms, maples	Verticillium wilt	*Verticillium albo-atrum*	General	Cultural
Sycamore	Anthracnose	*Gnomonia platoni*	General	Cultural, fungicide
Several	Adverse soil conditions, air pollution	Several	General	Cultural
Conifers				
Several	Adverse soil conditions, air pollution	Several	General	Cultural

TABLE A-2 Forest Diseases in the South (Gulf States, South Atlantic)

Host	Disease	Agent	Location	Control
Forest				
All conifers	Annosus root rot	*Fomes annosus*	General	Cultural
Shortleaf pine	Littleleaf disease	*Phytophthora cinnamomi* and environmental conditions	General	None
Longleaf pine	Brown spot disease	*Scirrhia acicola*	General	Fungicide in nursery, none in plantations
Loblolly and slash	Fusiform rust	*Cronartium fusiforme*	General	Fungicide in nursery, cultural or nothing in plantations
Loblolly pine	Comandra blister rust	*Cronartium commandrae*	Kentucky, Tennessee	Cultural
White pine	Blister rust	*Cronartium ribicola*	Virginia	Herbicide, ribes removal

TABLE A-2 (continued)

Host	Disease	Agent	Location	Control
<u>Urban</u>				
Oak	Oak-wilt	*Ceratocystis fagacearum*	North Carolina	None, eradication
Mimosa	Mimosa wilt	*Fusarium oxysporum* f. *perniciosum*	General	None, eradication
Elm	Dutch elm disease	*Ceratocystis ulmi*	General	Cultural, pesticides

TABLE A-3 Forest Diseases in the Pacific Coast Region (Pacific Northwest and Southwest)

Host	Disease	Agent	Location	Control
Lodgepole pine Ponderosa pine	Elytroderma needle blight	*Elytroderma deformans*	Intermountain, California	Cultural
Several conifers	Verticicladiella root disease	*Verticicladiella wageneri*	California, Southwest	Cultural
Lodgepole pine Douglas fir, other conifers	Dwarf mistletoe	*Arceuthobium* spp.	Intermountain, Southwest, Northwest	Cultural
Ponderosa pine	Red rot	*Polyporus anceps*	Southwest	Cultural
Several conifers	Annosus root rot	*Fomes annosus*	California, Intermountain	Cultural
Douglas fir	Rhizina root rot	*Rhizina undulata*	Washington	None
Douglas fir	Red ring rot	*Fomes pini*	Northwest	Cultural
Western hemlock, true firs	Brown stringy rot	*Echinodontium tinctorium*	Northwest	Cultural
Several conifers	Yellow ring rot	*Poria weirii*	Northwest	Cultural

TABLE A-4 Forest Diseases in the Rocky Mountains (Northern and Southern Rocky Mountains)

Host	Disease	Agent	Location	Control
Conifers	Dwarf mistletoe	*Arceuthobium* spp.	Northern, Central	Cultural
Conifers	Armillaria root rot	*Armillaria mellea*	Northern, Central	Cultural
Conifers	Yellow ring rot	*Poria weirii*	Northern	Cultural
Conifers	Red ring rot	*Fomes pini*	Northern	Cultural
Grand fir, western hemlock	Brown stringy rot	*Echinodontium tinctorium*	Northern	Cultural
Western white pine	Blister rust	*Cronartium ribicola*	Northern	Cultural
Lodgepole pine	Comandra rust	*Cronartium comandrae*	Northern	Cultural
Pinyon pine	Verticicladiella root disease	*Verticicladiella wageneri*	Southern	Cultural

APPENDIX B

REPRESENTATIVE INSECT PESTS IN FOREST AREAS OF THE UNITED STATES

TABLE B-1 Insect Pests of the North (Lake States, Central, Middle Atlantic, New England)

Host	Insect	Damage	Location	Control
Forest				
Balsam fir, spruce	Spruce budworm (*Choristoneura fumiferana*)	Defoliation	Maine, Lake states	Chemical, cultural
Jack pine	Jack pine budworm (*Choristoneura pinus*)	Defoliation	Lake states	Chemical, cultural (salvage)
Hemlock, balsam fir, spruce	Hemlock looper (*Lambdina fiscellaria*)	Defoliation	Eastern states	Chemical, cultural
Many pines (especially plantations)	Redheaded pine sawfly (*Neodiprion lecontei*) and other sawflies	Defoliation	Eastern states	Chemical, cultural, biological
White pine, jack pine, spruces	White pine weevil (*Pissodes strobi*)	Shoot damage	General	Chemical, cultural (mechanical)
Red pine, jack pine (especially plantations)	Saratoga and pine spittlebugs (*Aprophora* spp.)	Sapsucking	General	Chemical, cultural

True firs	Balsam woolly aphid (*Adelges piceae*)	Sapsucking	General	Cultural
Pines (especially seedlings)	Pales weevil (*Hylobius pales*)	Shoot damage (mortality in seedlings)	General	Cultural, chemical
Pines	Pine tip moths (*Rhyacionia* spp.)	Shoot damage (deformities)	General	Cultural
Larch type	Larch sawfly (*Pristiphora erichsonii*)	Defoliation	General	Chemical, cultural
Larch type	Larch casebearer (*Coleophora laricella*)	Defoliation	General	Biological, chemical
Aspens, maples	Forest tent caterpillar (*Malacosoma disstria*)	Defoliation	General	Chemical (high-value trees only)
Sugar maple, yellow birch, beech	Saddled prominent (*Heterocampa guttivitta*)	Defoliation	General	Chemical (high-value trees only)
<u>Urban</u>				
Oak, birches, other hardwoods (except tulip tree), hemlock	Gypsy moth (*Porthetria dispar*)	Defoliation	New England, Middle Atlantic	Chemical, cultural

TABLE B-1 (continued)

Host	Insect	Damage	Location	Control
Oaks	Red-humped oakworm (*Symmerista canicosta*)	Defoliation	Michigan, Iowa, Missouri	Chemical
Many shade trees	Aphids (many species of order Homoptera, family Aphididae)	Sapsucking; secreting honeydew	General	Chemical, cultural
Many shade trees	Scales (many species of order Homoptera, family Coccidae)	Sapsucking; some secrete honeydew	General	Cultural, chemical
Many shade trees	Mites (smaller members of order Acarina)	Sapsucking; some secrete honeydew defoliate	General	Chemical, cultural

TABLE B-2 Insect Pests of the South (Gulf States, South Atlantic)

Host	Insect	Damage	Location	Control
Southern pines	Southern pine beetle (*Dendroctonus frontalis*)	Phloem damage (mortality)	General	Cultural (salvage), chemical (lindane)
Pines (especially young trees)	Pine engraver beetles (*Ips* spp.)	Phloem damage (mortality)	Florida, North Carolina, Texas, Arkansas	Cultural, chemical (lindane)
Pines	Black turpentine beetle (*Dendroctonus terebrans*)	Phloem damage (mortality)	Virginia, Florida, North Carolina, Arkansas	Cultural (salvage), chemical (lindane)
Pines (especially plantations)	Sawflies, especially pine, redheaded, and Arkansas (*Neodiprion* spp.)	Defoliation	General	Chemical, cultural, biological
Oak, water tupelo	Forest tent caterpillar (*Malacosoma disstria*)	Defoliation	Alabama, Louisiana, Florida, Kentucky, Texas	Chemical (high-value trees only)
Pines (especially seedlings)	Pales weevil (*Hylobius pales*)	Shoot damage (mortality in seedlings)	General	Cultural, chemical

TABLE B-2 (continued)

Host	Insect	Damage	Location	Control
Pines (especially in seed orchards)	Pine tip moth (*Rhyacionia* spp.)	Shoot damage	Louisiana	Chemical (high-value trees only)
Oak	Variable oak leaf caterpillar (*Heterocampa manteo*)	Defoliation	Arkansas, northwest Louisiana	Chemical (high-value trees only)
Oak	Walkingstick (*Diapheromera femorata*)	Defoliation	Arkansas	Chemical (high-value trees only)

TABLE B-3 Insect Pests of Pacific Coast (Pacific Northwest and Southwest, excluding Hawaii)

Host	Insect	Damage	Location	Control
Douglas fir (especially damaged or overmature)	Douglas fir beetle (*Dendroctonus pseudotsugae*)	Phloem damage	Oregon, Washington	Cultural (salvage, sanitation), biological (pheromones--experimental)
Lodgepole pine, western white pine, ponderosa pine (especially damaged)	Mountain pine beetle (*D. ponderosae*)	Phloem damage	Eastern Oregon and Washington, California	Cultural (salvage)
Engelmann spruce, white spruce (damaged or mature)	Spruce beetle (*D. rufipennis*)	Phloem damage	Washington, Montana, Alaska	Cultural (salvage)
Ponderosa, Coulter pine (especially damaged)	Western pine beetle (*D. brevicomis*)	Phloem damage	California, Oregon, Washington	Cultural (salvage), biological (pheromones--experimental)

TABLE B-3 (continued)

Host	Insect	Damage	Location	Control
Douglas fir, true firs	Douglas fir tussock moth [*Orgyia* (*Hemerocampa*) *pseudotsugata*]	Defoliation	Oregon, Washington, California	Chemical, cultural (salvage), biological (experimental)
Spruce	Western spruce budworm (*Choristoneura occidentalis*)	Defoliation	Southern Oregon	Chemical
Hemlock	Western black-headed budworm (*Acleris gloverana*)	Defoliation	Washington, Alaska	Chemical
Larch type	Larch casebearer (*Coleophora laricella*)	Defoliation (little mortality)	Northeast Oregon, southeast Washington	Biological (parasites--experimental)
True firs	Fir engraver (*Scolytus ventralis*)	Phloem damage	Oregon, Washington, California	Cultural (salvage)
Pines (especially damaged)	Pine engraver beetles (*Ips* spp.)	Phloem damage	Oregon, Washington, California	Cultural, lindane (high-value trees only)
Hemlock, spruce	Hemlock sawfly (*Neodiprion tsugae*)	Defoliation	General	Chemical
Lodgepole pine	Lodgepole needle miner (*Coleotechnites milleri*)	Defoliation	California, central Oregon	Cultural

TABLE B-4 Insect Pests of the Rocky Mountains (Northern and Southern Rocky Mountains)

Host	Insect	Damage	Location	Control
Douglas fir, true fir, spruce	Western spruce budworm (*Choristoneura occidentalis*)	Defoliation	Idaho, Montana, Wyoming, Colorado	Chemical
Douglas fir, various pines	Douglas fir beetle (*Dendroctonus pseudotsugae*), mountain pine beetle (*D. ponderosae*), Ips engraver beetles (*Ips* spp.)	Phloem damage	General	Cultural (salvage)
Engelmann spruce	Spruce beetle (*Dendroctonus rufipinnis*)	Phloem damage	Wyoming, Utah, Colorado	Cultural (salvage)
Ponderosa pine (mature and overmature)	Roundheaded pine beetle (*D. adjunctus*)	Phloem damage	Nevada, New Mexico, Arizona	Cultural (salvage), chemical
Douglas fir, white fir, shade and ornamental trees	Douglas fir tussock moth (*Hemerocampa pseudotsugata*)	Defoliation	Nevada, Idaho, New Mexico	Chemical

TABLE B-4 (continued)

Host	Insect	Damage	Location	Control
Lodgepole pine	Lodgepole terminal weevil (*Pissodes terminalis*)	Shoot damage	Idaho, Utah, Colorado, Wyoming	
Grand fir, western red cedar, western hemlock, understory hemlock	Western hemlock looper (*Lambdina fiscellaria lugubrosa*)	Defoliation	Northern Idaho	Chemical
Ponderosa pine	Pine butterfly (*Neophasia menapia*)	Defoliation	Idaho, Montana	Chemical
Western larch	Larch casebearer (*Coleophora laricella*)	Defoliation	Northern Idaho	Biological, chemical
Aspens	Tent caterpillar (*Malacosoma* spp.)	Defoliation	Central and southern Rocky Mountains	Chemical, cultural
Pines	Western and southwestern pine tip moths (*Rhyacionia* spp.)	Shoot damage	Colorado	Biological, chemical

APPENDIX C

IMPORTANT VERTEBRATE PESTS IN FORESTS OF THE UNITED STATES

Common Name	Scientific Name
Shrew	*Sorex* spp.
Black bear	*Ursus americanus*
Mountain beaver	*Aplodontia rufa*
Chipmunk	*Eutamias* spp.
Tree squirrel	*Tamiasciurus* spp.
	Sciurus
Pocket gopher	*Thomomys* spp.
	Geomys spp.
Deer mouse	*Peromyscus* spp.
Woodrat	*Neotoma* spp.
Cotton rat	*Sigmodon hispidus*
Vole	*Microtus* spp.
	Clethrionomys spp.
Porcupine	*Erethizon dorsatum*
Hare	*Lepus americanus*
Rabbit	*Sylvilagus* spp.
Elk	*Cervus canadensis*
Mule deer	*Odocoileus hemionus*
White-tailed deer	*O. virginianus*

APPENDIX D

FOREST AREA (1,000 ACRES) AERIALLY SPRAYED WITH INSECTICIDES IN THE UNITED STATES, 1945-1974, BY YEAR AND TARGET SPECIES

Year	WB	GM	SB	DFTM	PB	PPL	FC	JPBW	WHL	SS	GBTC	PTM	ES	BHPS	WBHBH	RHPS	SP
1945	-	5	-	-	-	-	-	-	11	-	-	-	-	-	-	-	-
1946	-	61	-	-	-	-	-	-	-	3	-	-	-	-	-	-	-
1947	-	107	-	414	-	-	-	-	-	3	-	-	-	-	-	9	-
1948	4	211	-	-	-	-	-	-	-	1	-	-	-	-	-	2	-
1949	267	390	-	-	-	-	-	-	-	3	-	-	-	-	-	-	-
1950	938	583	-	-	-	-	-	-	-	11	16	-	-	-	-	-	-
1951	916	178	-	-	-	-	-	4	-	5	-	-	-	-	-	-	-
1952	669	202	-	-	-	-	-	-	-	4	-	-	-	-	-	-	-
1953	508	180	-	-	-	-	-	-	-	11	-	-	-	-	-	-	-
1954	75	1,372	20	-	255	-	-	-	-	8	3	-	-	-	-	-	-
1955	2,333	1,083	-	-	-	226	-	-	-	9	2	-	-	-	-	-	-
1956	1,367	926	-	10	-	-	-	-	-	3	-	-	-	-	-	-	-
1957	1,445	3,395	-	-	-	-	-	40	-	6	4	-	-	-	6	-	-
1958	936	516	314	-	-	-	-	-	-	4	3	-	-	-	-	-	-
1959	127	115	7	29	-	-	-	1	-	2	-	-	1	18	-	-	-
1960	119	65	242	13	-	-	-	1	-	3	-	-	1	-	-	2	-
1961	-	142	66	1	-	-	1	40	-	2	-	-	2	-	-	-	-
1962	1,027	427	57	-	-	-	2	37	33	4	-	26	9	-	-	-	-
1963	1,220	414	498	12	-	-	1	-	59	5	-	7	11	-	10	-	-
1964	684	135	58	-	-	-	-	-	-	-	3	8	-	-	-	-	-
1965	8	226	-	245	-	-	107	-	-	-	23	-	-	-	-	-	-
1966	148	275	-	-	-	-	1	-	-	-	-	-	-	-	-	-	-
1967	2	203	101	-	-	-	-	-	-	1	-	-	-	-	-	-	-
1968	6	231	10	-	-	-	8	-	1	3	-	-	-	-	-	-	-
1969	8	91	18	-	-	-	10	-	-	6	-	-	-	-	-	1	-
1970	-	223	212	-	-	-	2	-	-	5	-	-	-	-	-	-	14
1971	9	391	10	-	-	-	-	-	-	-	-	-	-	-	-	-	-
1972	-	177	500	-	-	-	-	-	-	-	-	-	-	-	-	-	-
1973	-	189	473	70	-	-	-	-	-	-	-	-	-	-	-	-	-
1974	-	171	434	466	-	-	-	-	-	-	-	-	-	-	-	-	-
	12,816	12,684	3,020	1,260	255	239	132	123	104	102	54	41	24	18	16	14	14

INSECT ABBREVIATIONS FOR APPENDIX D

WB	WESTERN BUDWORM, *Choristoneura occidentalis*
GM	GYPSY MOTH, *Porthretria dispar*
SB	SPRUCE BUDWORM, *Choristoneura fumiferana*
DFTM	DOUGLAS-FIR TUSSOCK MOTH, *Orgyia (=Hemerocampa) pseudotsugata*
PB	PINE BUTTERFLY, *Neophasia menapia*
PPL	PITCH-PINE LOOPER, *Lambdina athasaria pellucidaria*
FC	FALL CANKERWORM, *Alsophila pometaria*
JPBW	JACK-PINE BUDWORM, *Choristoneura pinus*
WHL	WESTERN HEMLOCK LOOPER, *Lambdina fiscellaria lugubrosa*
SS	SARATOGA SPITTLEBUG, *Aphrophora saratogensis*
GBTC	GREAT BASIN TENT CATERPILLAR, *Malacosoma fragile*
PTM	PINE TUSSOCK MOTH, *Dasychira plagiata*
ES	ELM SPANWORM, *Ennomos subsignarius*
BHPS	BLACKHEADED PINE SAWFLY, *Neodiprion excitans*
WBHBW	WESTERN BLACKHEADED BUDWORM, *Acleris gloverana*
RHPS	REDHEADED PINE SAWFLY, *Neodiprion lecontei*
SP	SADDLED PROMINENT, *Heterocampa guttivitta*

ear	LPNM	NMFL	PRW	WFNM	GH	WFS	AS	EHL	FW	FTC	LS	LLPS	LPPS	SWPTM	UC	VPS	MB	Total
945	-	-	-	-	-	-	-	-	-	-	-	-	-	-	-	-	-	16
946	-	-	-	-	-	-	-	-	-	-	-	-	-	-	-	-	-	64
947	-	-	-	-	-	-	-	-	-	-	-	-	-	-	-	-	-	533
948	-	-	-	-	-	-	-	-	-	-	-	-	-	-	-	-	-	218
949	-	-	-	1	-	-	-	-	-	-	-	-	-	-	-	-	-	661
950	-	-	-	-	-	-	-	-	-	-	-	-	-	-	-	-	-	1,548
951	-	-	-	-	-	-	2	-	-	-	-	-	-	-	-	-	-	1,105
952	-	10	3	-	-	-	-	-	-	1	-	-	-	-	-	-	-	889
953	-	-	-	-	-	-	-	-	-	-	-	-	1	-	-	-	-	700
954	-	-	-	-	-	-	-	-	-	-	-	-	-	-	-	-	-	1,733
955	-	-	-	-	-	-	-	-	-	-	-	-	-	-	-	-	-	3,653
956	-	-	-	-	-	-	-	-	-	-	-	-	-	-	-	-	-	2,306
957	-	-	1	4	-	-	-	-	-	-	-	1	-	-	-	1	-	4,902
958	-	-	-	-	-	-	-	-	-	-	-	-	-	-	-	-	-	1,774
959	3	-	2	-	-	-	-	-	-	-	-	-	-	-	-	-	-	305
960	-	-	-	-	-	-	-	-	-	-	-	-	-	-	-	-	-	446
961	5	-	-	-	-	-	-	-	-	-	-	-	-	-	-	-	-	259
962	-	-	-	-	-	-	-	-	1	-	-	-	-	-	-	-	-	1,623
963	5	-	-	-	-	-	-	-	-	-	-	-	-	-	1	-	-	2,243
964	-	-	-	-	-	3	-	-	-	-	-	-	-	-	-	-	-	891
965	-	-	-	-	-	-	-	1	-	-	-	-	-	-	-	-	-	610
966	-	-	-	-	-	-	-	-	-	-	-	-	-	-	-	-	-	424
967	-	-	-	-	-	-	-	-	-	-	-	-	-	-	-	-	-	307
968	-	-	-	-	4	-	-	-	-	-	1	-	-	1	-	-	-	265
969	-	-	-	-	-	-	-	-	-	-	-	-	-	-	-	-	-	134
970	-	-	-	-	-	-	-	-	-	-	-	-	-	-	-	-	-	456
971	-	-	-	-	-	-	-	-	-	-	-	-	-	-	-	-	-	410
972	-	-	-	-	-	-	-	-	-	-	-	-	-	-	-	-	-	690
973	-	-	-	-	-	-	-	-	-	-	-	-	-	-	-	-	-	732
974	-	-	-	-	-	-	-	-	-	-	-	-	-	-	-	-	6	1,077
	14	13	10	6	5	4	3	2	1	1	1	1	1	1	1	1	6	30,974

LPNM LODGEPOLE NEEDLE MINER, *Coleotechnites milleri*
NMFL NEW MEXICO FIR LOOPER, *Galenara consimilis*
PRW PINE REPRODUCTION WEEVIL, *Cylindrocopturus eatoni*
WFNM WHITE FIR NEEDLE MINER, *Epinotia meritana*
GH GRASSHOPPER, *Bradynotes obesa opima*
WFS WHITE FIR SAWFLY, *Neodiprion abietis*
AS ARKANSAS SAWFLY, *Neodiprion taedae linearis*
EHL EASTERN HEMLOCK LOOPER, *Lambdina fiscellaria fiscellaria*
FW FALL WEBWORM, *Hyphantria cunea*
FTC FOREST TENT CATERPILLAR, *Malacosoma disstria*
LS LARCH SAWFLY, *Pristiphora erichsonii*
LLPS LOBLOLLY PINE SAWFLY, *Neodiprion taedae taedae*
LPPS LODGEPOLE PINE SAWFLY, *Neodiprion burkei*
SWPTM SOUTHWESTERN PINE TIP MOTH, *Rhyacionia neomexicana*
UC UTAH CANKERWORMS
VPS VIRGINIA PINE SAWFLY, *Neodiprion pratti pratti*
MB MODOC BUDWORM, *Choristoneura virdis*

APPENDIX E

FOREST AREA (1,000 ACRES) AERIALLY SPRAYED WITH INSECTICIDES IN THE UNITED STATES, 1945-1974, BY YEAR AND MATERIAL USED[a]

Year	DDT	Car-baryl	Mexacar-bamate	Mala-thion	Fenitro-thion	Trichlor-fon	DFTM Virus	Gar-dona	Lead Arsenate	Methoxy-chlor	Phospham-idon	Bacil-lus	Dime-thoate	Pro-poxur	GBTC Virus	Total
1945	7	--	--	--	--	--	--	--	9	--	--	--	--	--	--	16
1946	64	--	--	--	--	--	--	--	--	--	--	--	--	--	--	64
1947	533	--	--	--	--	--	--	--	--	--	--	--	--	--	--	533
1948	218	--	--	--	--	--	--	--	--	--	--	--	--	--	--	218
1949	661	--	--	--	--	--	--	--	--	--	--	--	--	--	--	661
1950	1,548	--	--	--	--	--	--	--	--	--	--	--	--	--	--	1,548
1951	1,105	--	--	--	--	--	--	--	--	--	--	--	--	--	--	1,105
1952	880	--	--	--	--	--	--	--	--	--	--	--	--	--	--	880
1953	700	--	--	--	--	--	--	--	--	--	--	--	--	--	--	700
1954	1,733	--	--	--	--	--	--	--	--	--	--	--	--	--	--	1,733
1955	3,653	--	--	--	--	--	--	--	--	--	--	--	--	--	--	3,653
1956	2,306	--	--	--	--	--	--	--	--	--	--	--	--	--	--	2,306
1957	4,898	--	--	4	--	--	--	--	--	--	--	--	--	--	--	4,902
1958	1,773	--	--	--	--	--	--	--	--	--	--	--	--	--	1	1,774
1959	216	86	--	3	--	--	--	--	--	--	--	--	--	--	--	305
1960	430	11	--	1	--	--	4	--	--	--	--	--	--	--	--	446
1961	218	29	--	5	--	--	--	--	--	7	--	--	--	--	--	259
1962	1,592	31	--	--	--	--	--	--	--	--	--	--	--	--	--	1,623
1963	2,074	130	--	20	--	--	12	--	--	--	7	--	1	--	--	2,243
1964	632	97	--	161	--	--	--	--	--	--	--	2	--	--	--	891
1965	363	234	--	11	--	--	--	--	--	--	--	1	--	--	--	610
1966	--	275	5	143	--	--	--	--	--	--	--	--	--	--	--	424
1967	100	203	3	1	--	--	--	--	--	--	--	--	1	--	--	307
1968	--	201	7	7	10	39	--	--	--	--	--	--	--	1	--	265
1969	--	95	16	6	--	6	--	10	--	--	--	--	--	--	--	134
1970	--	233	--	5	212	3	--	3	--	--	--	--	--	--	--	456
1971	--	390	19	--	--	--	--	1	--	--	--	--	--	--	--	410
1972	--	172	500	13	--	3	--	--	--	--	--	2	--	--	--	690
1973	--	61	543	--	--	125	--	--	--	--	--	3	--	--	--	732
1974	425	90	470	--	--	77	--	--	--	--	--	15	--	--	--	1,077
	26,138	2,338	1,563	380	222	253	16	14	9	7	7	23	2	1	1	30,974

[a]Compiled by Charles Sartwell, USFS, Corvallis, Oregon, and David Alligood, USFS, Washington, D.C.